Glencoe McGraw-Hill

Grade
8

Math
Triumphs

Book 2:
Geometry and
Measurement

Authors

Basich Whitney • Brown • Dawson • Gonsalves • Silbey • Vielhaber

McGraw Hill Glencoe

Photo Credits

All coins photographed by United States Mint.
All bills photographed by Michael Houghton/StudioOhio.
Cover Jupiterimages; **iv** (1 7 8)File Photo, (2 3)The McGraw-Hill Companies,
(4 5 6)Doug Martin; **vi** C. Borland/Getty Images; **vii** CORBIS; **viii** PunchStock;
136–137 CORBIS; **142** Getty Images; **149** (t)Alamy, (b)Getty Images; **150** Peter
Adams/CORBIS; **151, 158** CORBIS; **159** PunchStock; **178** SuperStock;
183, 184 CORBIS; **185** PunchStock; **189** John Flournoy/The McGraw-Hill
Companies; **190** (t)Steve Cohen/Jupiterimages, (bl)PunchStock, (br)Getty Images;
192 PunchStock; **198** CORBIS; **199** SuperStock; **204** Erik Dreyer/Getty Images;
206 Ken Cavanagh/The McGraw-Hill Companies; **207** Ryan McVay/Getty Images;
208 Charles Smith/CORBIS; **216–217** Jupiterimages; **221, 224** CORBIS; **228** Age
Fotostock; **230** PunchStock; **237** D. Normark/Getty Images; **245** Jupiterimages;
246 Robert Glusic/Getty Images; **254** John Flournoy/The McGraw-Hill Companies.

CURRICULUM RESOURCES CENTER
MILNE LIBRARY
STATE UNIVERSITY COLLEGE
1 COLLEGE CIRCLE
GENESEO NY 14454 LIBRARY
WITHDRAWN
S.U.N.Y GENESEO

Copyright © Glencoe/McGraw-Hill, a division of The McGraw-Hill Companies, Inc.

The McGraw-Hill Companies

 Macmillan/McGraw-Hill
Glencoe

Copyright © 2009 The McGraw-Hill Companies, Inc. All rights reserved. No part of this publication
may be reproduced or distributed in any form or by any means, or stored in a database or retrieval
system, without the prior written consent of The McGraw-Hill Companies, Inc., including, but not
limited to, network storage or transmission, or broadcast for distance learning.

Send all inquiries to:
Glencoe/McGraw-Hill
8787 Orion Place
Columbus, OH 43240-4027

ISBN: 978-0-07-888214-2
MHID: 0-07-888214-1

Printed in the United States of America.

3 4 5 6 7 8 9 10 066 17 16 15 14 13 12 11 10 09

Math Triumphs
Grade 8, Book 2

Math Triumphs

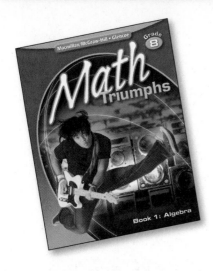

Book 1

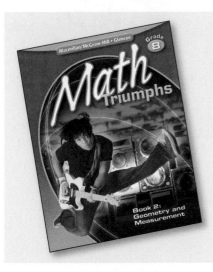

Book 2

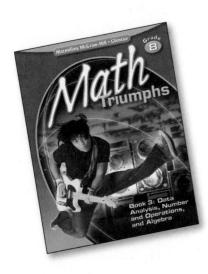

Book 3

Copyright © Glencoe/McGraw-Hill, a division of The McGraw-Hill Companies, Inc.

Authors and Consultants

AUTHORS

Frances Basich Whitney
Project Director, Mathematics K–12
Santa Cruz County Office of Education
Capitola, California

Kathleen M. Brown
Math Curriculum Staff Developer
Washington Middle School
Long Beach, California

Dixie Dawson
Math Curriculum Leader
Long Beach Unified
Long Beach, California

Philip Gonsalves
Mathematics Coordinator
Alameda County Office of Education
Hayward, California

Robyn Silbey
Math Specialist
Montgomery County Public Schools
Gaithersburg, Maryland

Kathy Vielhaber
Mathematics Consultant
St. Louis, Missouri

CONTRIBUTING AUTHORS

Viken Hovsepian
Professor of Mathematics
Rio Hondo College
Whittier, California

FOLDABLES Study Organizer **Dinah Zike**
Educational Consultant,
Dinah-Might Activities, Inc.
San Antonio, Texas

CONSULTANTS

Assessment

Donna M. Kopenski, Ed.D.
Math Coordinator K–5
City Heights Educational Collaborative
San Diego, California

Instructional Planning and Support

Beatrice Luchin
Mathematics Consultant
League City, Texas

ELL Support and Vocabulary

ReLeah Cossett Lent
Author/Educational Consultant
Alford, Florida

Copyright © Glencoe/McGraw-Hill, a division of The McGraw-Hill Companies, Inc.

Reviewers

Each person below reviewed at least two chapters of the Student Edition, providing feedback and suggestions for improving the effectiveness of the mathematics instruction.

Patricia Allanson
Mathematics Teacher
Deltona Middle School
Deltona, Florida

Debra Allred
Sixth Grade Math Teacher
Wiley Middle School
Leander, Texas

April Chauvette
Secondary Mathematics Facilitator
Leander Independent School District
Leander, Texas

Amy L. Chazarreta
Math Teacher
Wayside Middle School
Fort Worth, Texas

Jeff Denney
Seventh Grade Math Teacher, Mathematics
  Department Chair
Oak Mountain Middle School
Birmingham, Alabama

Franco A. DiPasqua
Director of K-12 Mathematics
West Seneca Central
West Seneca, New York

David E. Ewing
Teacher
Bellview Middle School
Pensacola, Florida

Mark J. Forzley
Eighth Grade Math Teacher
Westmont Junior High School
Westmont, Illinois

Virginia Granstrand Harrell
Education Consultant
Tampa, Florida

Russ Lush
Sixth Grade Math Teacher, Mathematics
  Department Chair
New Augusta - North
Indianapolis, Indiana

Joyce B. McClain
Middle School Math Consultant
Hillsborough County Schools
Tampa, Florida

Suzanne D. Obuchowski
Math Teacher
Proctor School
Topsfield, Massachusetts

Karen L. Reed
Sixth Grade Pre-AP Math
Keller ISD
Keller, Texas

Deborah Todd
Sixth Grade Math Teacher
Francis Bradley Middle School
Huntersville, North Carolina

Susan S. Wesson
Teacher (retired)
Pilot Butte Middle School
Bend, Oregon

Copyright © Glencoe/McGraw-Hill, a division of The McGraw-Hill Companies, Inc.

Contents

Chapter 4 — Angle Measures

Mystic, Connecticut

Copyright © Glencoe/McGraw-Hill, a division of The McGraw-Hill Companies, Inc.

Chapter 5 Ratios, Rates, and Similarity

Hill, a division of The McGraw-Hill Companies, Inc.

Stowe, Vermont

Contents

Copyright © Glencoe/McGraw-Hill, a division of Th

Chapter 6

Squares, Square Roots, and the Pythagorean Theorem

Miami, Florida

SCAVENGER HUNT

Let's Get Started

Use the Scavenger Hunt below to learn where things are located in each chapter.

1. What is the title of Lesson 5-2?

2. What is the Key Concept of Lesson 4-4?

3. On what page can you find the vocabulary term *complimentary angles* in Lesson 4-3?

4. What are the vocabulary words for Lesson 5-4?

5. How many exercises are presented in the Vocabulary and Concept Check section of the Chapter 5 Study Guide?

6. What strategy is used in the Step-by-Step Problem-Solving Practice on page 190?

7. List the first problem-solving step mentioned in Exercise 9 on page 245.

8. Describe the photo that accompanies Exercise 8 on page 149.

9. On what pages will you find the Test Practice for Chapter 6?

10. In Chapter 5, find the logo and internet address that tells you where you can take the Online Readiness Quiz.

Copyright © Glencoe/McGraw-Hill, a division of The McGraw-Hill Companies, Inc.

Angle Measures

Angle measures are important in building and design.

We must understand and measure angles to build planes, buildings, maps, rockets, cars, and bicycles.

Copyright © Glencoe/McGraw-Hill, a division of The McGraw-Hill Companies, Inc.

STEP 1 Quiz

Math Online › Are you ready for Chapter 4? Take the Online Readiness Quiz at *glencoe.com* to find out.

STEP 2 Preview

Get ready for Chapter 4. Review these skills and compare them with what you will learn in this chapter.

What You Know	What You Will Learn
You know that if you do a 180° turn, you turn to face the opposite way. 	*Lesson 4-1* Angles are figures that are often measured in degrees. 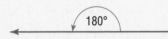 This angle measures 180°. It is a straight angle.
You know how to describe and recognize some figures. **TRY IT!** Describe the figure. **1** name of figure _____ **2** number of sides _____ **3** number of angles _____	*Lesson 4-2* Triangles can be classified by their angles and sides. An equilateral triangle has three sides that are equal in length. All three angles in an acute triangle are greater than 0° and less than 90°.
You know about lines, line segments, and rays. **TRY IT!** Name each figure below. **4** ⟶ _____ **5** ⟷ _____ **6** •—• _____	*Lesson 4-4* When a **transversal** passes through two parallel lines, the angles created have special relationships to one another. ∠A and ∠B are **corresponding angles**. These angles are in the same position on the two lines in relation to the transversal.

Copyright © Glencoe/McGraw-Hill, a division of The McGraw-Hill Companies, Inc.

Angles

KEY Concept

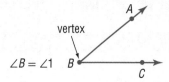

vertex

$\angle B = \angle 1$ B

A

C

An **angle** can be named by using a point on one ray, the vertex, and then a point on the other ray, such as $\angle ABC$ or $\angle CBA$. An angle can also be named using the letter of the vertex, such as $\angle B$. An angle is sometimes named by a number, such as $\angle 1$.

Angles are often measured in **degrees**. They can be classified, or grouped, according to their measures. You can use a **protractor** to measure an angle.

The pink square inside $\angle DEF$ indicates that it is a right angle.

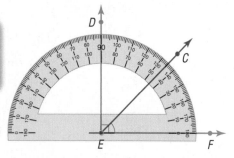

Acute angles measure between 0° and 90°. $\angle CEF$ is an acute angle with a measure of 45°.

Right angles measure exactly 90°. $\angle DEF$ is a right angle with a measure of exactly 90°.

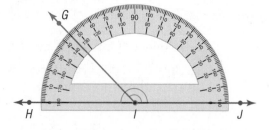

Obtuse angles measure between 90° and 180°. $\angle GIJ$ is an obtuse angle with a measure of 135°.

Straight angles measure exactly 180°. $\angle HIJ$ is a straight angle with a measure of exactly 180°.

VOCABULARY

angle
two rays with a common endpoint form an angle

degree
the most common unit of measure for angles

protractor
an instrument marked in degrees, used for measuring or drawing angles

ray
a part of a line that has one endpoint and extends indefinitely in one direction

vertex
the common endpoint of the two rays that form an angle (the plural is *vertices*)

Classify an angle as acute or obtuse to help you decide which scale of the protractor, the inner or outer, gives the angle's measurement.

Copyright © Glencoe/McGraw-Hill, a division of The McGraw-Hill Companies, Inc.

Example 1

Draw ∠DEF that measures 135°.

1. Draw $\overrightarrow{EF}$.

2. Place the center of the protractor at point E. Line up $\overrightarrow{EF}$ with the 0° mark on the protractor.

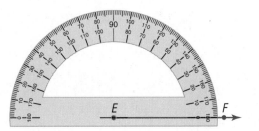

3. Use the inner scale. Draw a point D at 135°.

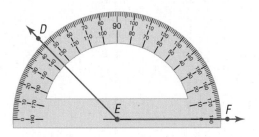

4. Draw $\overrightarrow{ED}$.

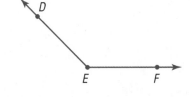

YOUR TURN!

Draw ∠WXY that measures 75°.

1. Draw $\overrightarrow{XY}$.

2. Place the center of the protractor at point _____. Line up $\overrightarrow{XY}$ with the _____° mark on the protractor.

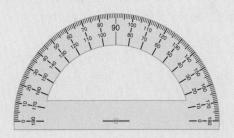

3. Use the inner scale. Draw a point W at _____°.

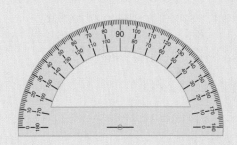

4. Draw $\overrightarrow{XW}$.

Copyright © Glencoe/McGraw-Hill, a division of The McGraw-Hill Companies, Inc.

GO ON

Example 2

Measure and identify the angle.

1. Place the center (hole or ⊥ symbol) of the protractor at the vertex, point Y.

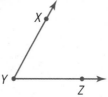

2. Line up $\overrightarrow{YZ}$ with the line that extends from 0° to 180° on the protractor.

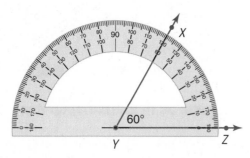

3. Look at point X. Use the 0° reading to know whether to read the inner scale or the outer scale on the protractor. Read the measure of the angle where $\overrightarrow{YX}$ passes through the inner scale.

∠XYZ measures 60°.

∠XYZ is an acute angle.

YOUR TURN!

Measure and identify the angle.

1. Place the center of the protractor at the vertex.

2. Line up $\overrightarrow{BC}$ with the line that extends from 0° to 180° on the protractor.

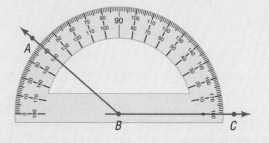

3. Look at point A. Use the 0° reading to know whether to read the inner scale or the outer scale on the protractor. Read the measure of the angle where $\overrightarrow{BA}$ passes through the inner scale.

∠ABC measures _____.

∠ABC is a(n) _____ angle.

Copyright © Glencoe/McGraw-Hill, a division of The McGraw-Hill Companies, Inc.

Who is Correct?

What is the measure of ∠B?

Hannah

115°

Tom

65°

Lamar

70°

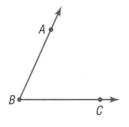

Circle correct answer(s). Cross out incorrect answer(s).

▶ Guided Practice

Draw an angle with the given measurement.

1 90°

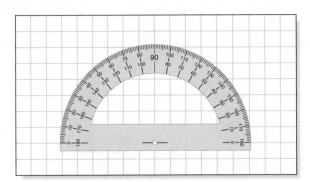

2 130°

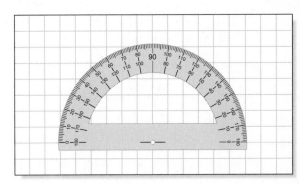

Step by Step Practice

3 Measure and identify the angle.

Step 1 Place the center of the protractor at point _____.

Step 2 Line up the vertex and the line on the protractor that extends from 0° to 180° with _____.

Step 3 Read from the _____ scale.

Step 4 Read the measure of the angle where _____ passes through the outer scale.

∠FGH measures _____.

∠FGH is a(n) _____ angle.

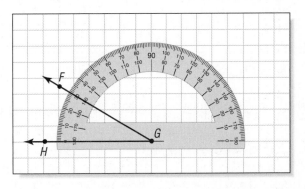

Measure and identify each angle.

4 ∠MNO measures _____.

∠MNO is a(n) _____ angle.

5 ∠TUV measures _____.

∠TUV is a(n) _____ angle.

Copyright © Glencoe/McGraw-Hill, a division of The McGraw-Hill Companies, Inc.

GO ON

Step by Step Problem-Solving Practice

Solve.

Problem-Solving Strategies
- ☑ Use a diagram.
- ☐ Look for a pattern.
- ☐ Guess and check.
- ☐ Solve a simpler problem.
- ☐ Work backward.

6 **KITES** On a windy April afternoon, Karl is flying kites with his younger brother. He notices that the corners of his kite form angles. What types of angles are formed by the corners of his kite?

Understand Read the problem. Write what you know.

Karl is looking at the angles formed

by _____ of his kite.

Plan Pick a strategy. One strategy is to use a diagram.

Solve Place the center of the protractor on the vertex labeled *A*. Line up the vertex and the line of the protractor that extends from 0° to 180°.

Read the measure of each angle.

Angle *A* = _____   Angle *B* = _____

Angle *C* = _____   Angle *D* = _____

The kite has _____ angles and _____ angles.

Check Review the definitions of acute and obtuse angles.

7 **ROAD SIGNS** Brenda's older brother is learning to drive. He must review the road signs for his driver's test. Describe the angles of the stop sign.

Check off each step.

_____ Understand: I underlined key words.

_____ Plan: To solve this problem, I will _____.

_____ Solve: The answer is _____.

_____ Check: I checked my answer by _____.

Copyright © Glencoe/McGraw-Hill, a division of The McGraw-Hill Companies, Inc.

8 **RAMPS** Refer to the diagram at the right. Describe the angles of this wheelchair ramp.

_____

9 [Reflect] Jacqueline says the measure of this acute angle is 80°. Explain her error.

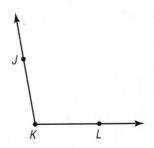

_____

_____

_____

 # Skills, Concepts, and Problem Solving

Draw an angle with the given measurement.

10 25°

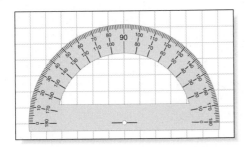

11 105°

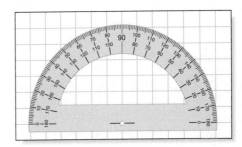

12 165°

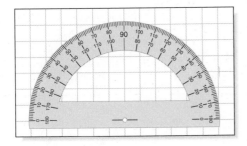

13 90°

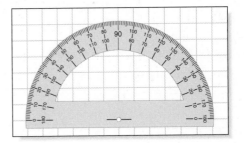

Copyright © Glencoe/McGraw-Hill, a division of The McGraw-Hill Companies, Inc.

GO ON

Measure and identify each angle.

14 ∠WXY measures _____.

∠WXY is a(n) _____ angle.

15 ∠PQR measures _____.

∠PQR is a(n) _____ angle.

Solve.

16 **CLOCKS**   Val's soccer practice starts at 4:45 P.M. What type of angle is formed by the hands of the clock at 4:45?

_____

17 **INTERIOR DESIGN**   Garrett measured the angle of two walls in his living room. The angle measured 90°. What type of angle is this?

_____

Vocabulary Check   **Write the vocabulary word that completes each sentence.**

18 A(n) _____ is formed by two rays with a common endpoint.

19 A(n) _____ is an instrument, marked in degrees, used for measuring or drawing angles.

20 A(n) _____ angle is an angle that measures greater than 90° but less than 180°.

21 A(n) _____ angle is an angle that measures greater than 0° but less than 90°.

22 **Writing in Math**   Explain how to draw ∠XYZ measuring 125°.

_____

_____

_____

Copyright © Glencoe/McGraw-Hill, a division of The McGraw-Hill Companies, Inc.

Lesson 4-2 Triangles

Triangles

KEY Concept

Triangles have three **sides** and three angles. Triangles can be classified by the lengths of their sides and by the measures of their angles.

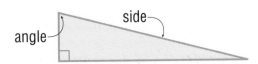

angle — side

Classify Triangles by Sides		
Example	Type	Number of Congruent Sides
	scalene triangle	0
	isosceles triangle	2
	equilateral triangle	3

Classify Triangles by Angles		
Example	Type	Measure of Angles
	acute triangle	all three angles measure less than 90°
	right triangle	one angle measures 90°
	obtuse triangle	one angle measures greater than 90°

VOCABULARY

acute angle
an angle with a measure greater than 0° and less than 90°

congruent
line segments that have the same length, or angles that have the same measure

obtuse angle
an angle with a measure greater than 90°, but less than 180°

right angle
an angle that measures 90°

side
a ray that is part of an angle

triangle
a polygon with three sides and three angles

Figures often have symbols that indicate congruence or right angles.

congruent sides

congruent angles

right angle

Copyright © Glencoe/McGraw-Hill, a division of The McGraw-Hill Companies, Inc.

Example 1

Classify triangle *ABC* by the lengths of its sides.

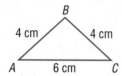

1. Find the measure of each side.

   $\overline{AB} = 4$ cm

   $\overline{BC} = 4$ cm

   $\overline{CA} = 6$ cm

2. Triangle *ABC* has 2 congruent sides.

3. Triangle *ABC* is an isosceles triangle.

YOUR TURN!

Classify triangle *DEF* by the lengths of its sides.

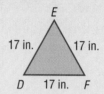

1. Find the measure of each side.

   $\overline{DE} = $ _____

   $\overline{EF} = $ _____

   $\overline{FD} = $ _____

2. Triangle *DEF* has _____ congruent sides.

3. Triangle *DEF* is a(n) _____ triangle.

Example 2

Classify triangle *GHI* by the measures of its angles.

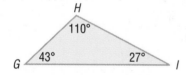

1. Find the measure of each angle.

   $m\angle G = 43°$

   $m\angle H = 110°$

   $m\angle I = 27°$

   > The abbreviation *m* means "The measure of."

2. Describe each angle.

   $\angle G$ is an acute angle.

   $\angle H$ is an obtuse angle.

   $\angle I$ is an acute angle.

3. Triangle *GHI* is an obtuse triangle.

YOUR TURN!

Classify triangle *JKL* by the measures of its angles.

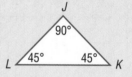

1. Find the measure of each angle.

   $m\angle J = $ _____

   $m\angle K = $ _____

   $m\angle L = $ _____

2. Describe each angle.

   $\angle J$ is _____ angle.

   $\angle K$ is _____ angle.

   $\angle L$ is _____ angle.

3. Triangle *JKL* is a(n) _____ triangle.

Copyright © Glencoe/McGraw-Hill, a division of The McGraw-Hill Companies, Inc.

Who is Correct?

Sketch an equilateral triangle.

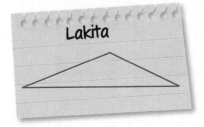

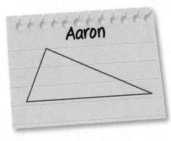

Circle correct answer(s). Cross out incorrect answer(s).

▶ Guided Practice

Classify each triangle by the lengths of its sides.

I

12 ft   12 ft

10 ft

The triangle is a(n)

_____.

2

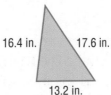

16.4 in.   17.6 in.

13.2 in.

The triangle is a(n)

_____ triangle.

Step by Step Practice

3 Sketch a right, scalene triangle.

Step 1 Use the protractor to make a right angle. Label the angle *ABC*, with _____ as the vertex.

Step 2 Measure the lengths of rays $\overrightarrow{BA}$ and $\overrightarrow{BC}$. The two rays _____ equal in length.

Step 3 Connect Point _____ and Point _____. The length of line segment $\overline{AC}$ _____ equal to the length of _____.

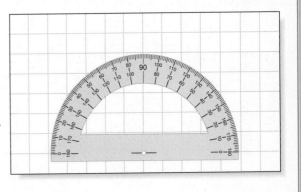

GO ON

Copyright © Glencoe/McGraw-Hill, a division of The McGraw-Hill Companies, Inc.

Classify each triangle by the measures of its angles.

4

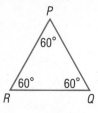

Angle	Measure	Type
∠P		
∠Q		
∠R		

Triangle *PQR* is a(n) _____ triangle.

5

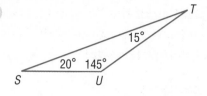

Angle	Measure	Type
∠S		
∠T		
∠U		

Triangle *STU* is a(n) _____ triangle.

Step by Step Problem-Solving Practice

Solve.

6 **CONSTRUCTION**   Mr. Richardson is constructing the frame for the roof of a building. The triangle below shows the lengths of the sides and the measures of the angles. Classify this triangle by its sides and its angles.

Problem-Solving Strategies

✓ Use a diagram.
☐ Look for a pattern.
☐ Guess and check.
☐ Act it out.
☐ Work backward.

Understand   Read the problem. Write what you know.

The frame of the roof is a _____.

Plan   Pick a strategy. One strategy is to use a diagram.

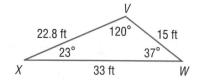

Solve   List the lengths of the sides and the measures of the angles.

$\overline{VW}$ = _____   *m*∠*V* = _____

$\overline{WX}$ = _____   *m*∠*W* = _____

$\overline{XV}$ = _____   *m*∠*X* = _____

This triangle has _____ congruent sides.

This triangle has _____ angles.

This triangle is a(n) _____ and a(n) _____ triangle.

Check   Review the definitions of the figure that you named.

Copyright © Glencoe/McGraw-Hill, a division of The McGraw-Hill Companies, Inc.

Solve.

7 **FLAGS** The flag of the Philippines is shown at the right. Classify the triangle portion by the length of its sides and the measure of its angles.

Check off each step.

_____ Understand: I underlined key words.

_____ Plan: To solve this problem, I will _____.

_____ Solve: The answer is _____.

_____ Check: I checked my answer by _____.

8 **TILES** Mrs. Jennings is tiling her bathroom. She needs to cut a square tile in half. Classify the triangle that will be created by the length of its sides and the measure of its angles.

_____

9 **Reflect** Can an equilateral triangle be classified as a right triangle? Explain.

_____

_____

_____

Sketch a figure with the description given.

10 obtuse, isosceles triangle

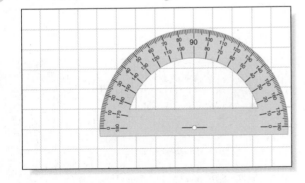

11 acute, equilateral triangle

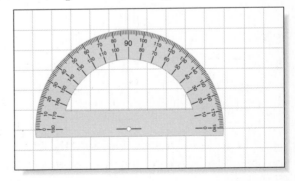

Copyright © Glencoe/McGraw-Hill, a division of The McGraw-Hill Companies, Inc.

GO ON

Skills, Concepts, and Problem Solving

Classify each figure by the lengths of its sides and the measures of its angles.

12

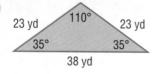

The figure is a(n)

_____.

13

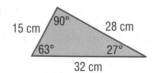

The figure is a(n)

_____.

14

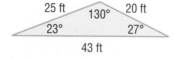

The figure is a(n)

_____.

15

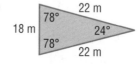

The figure is a(n)

_____.

Sketch a figure with the description given.

16 right, isosceles triangle

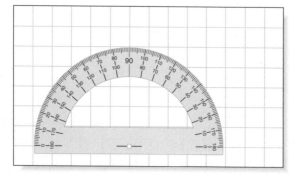

17 scalene, obtuse triangle

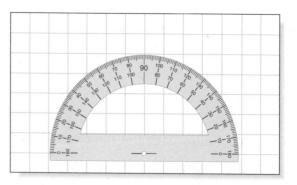

Solve.

18 **ART**   Gustavo made a model of the Great Pyramid in Egypt. Each triangle that made his pyramid had sides of length 30 centimeters, 26 centimeters, and 30 centimeters. Classify the triangle according to the measures of its sides.

_____

Copyright © Glencoe/McGraw-Hill, a division of The McGraw-Hill Companies, Inc.

19 LOGOS Trifecta Insurance Company uses the logo on the right. Classify the triangle in the logo by the length of its sides and the measure of its angles.

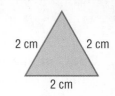

2 cm   2 cm

2 cm

_____

Vocabulary Check **Write the vocabulary word that completes each sentence.**

20 A(n) _____ is a triangle with at least two sides of the same length.

21 A(n) _____ is a triangle with one angle greater than 90° and less than 180°.

22 Writing in Math Explain how to classify a triangle.

_____

_____

▶ Spiral Review

Measure and identify each angle. (Lesson 4-1, p. 138)

23

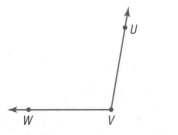

∠RST measures _____.

∠RST is a(n) _____ angle.

24

U

W   V

∠UVW measures _____.

∠UVW is a(n) _____ angle.

25 ARCHITECTURE Presently, the famed Tower of Pisa in Italy is leaning at about a 6° angle. This makes the angle of the leaning side to the ground measure about 84°. What type of an angle is the angle of the leaning side to the ground? (Lesson 4-1, p. 138)

ARCHITECTURE
Tower of Pisa

_____

Copyright © Glencoe/McGraw-Hill, a division of The McGraw-Hill Companies, Inc.

Draw an angle with the given measurement.

1 150°

2 40°

Measure and identify each angle.

3

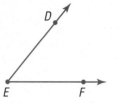

∠DEF measures _____.

∠DEF is a(n) _____ angle.

4

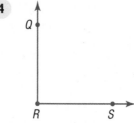

∠QRS measures _____.

∠QRS is a(n) _____ angle.

Classify each figure by the lengths of its sides and the measures of its angles.

5

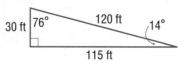

_____

6

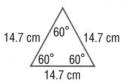

_____

Sketch a figure with the description given.

7 acute, scalene triangle

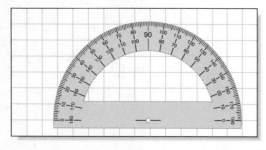

8 obtuse, equilateral triangle

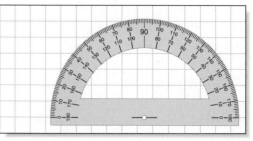

Solve.

9 **TIME** What type of angle is formed by the hands of the clock at 3:40?

_____

Copyright © Glencoe/McGraw-Hill, a division of The McGraw-Hill Companies, Inc.

Add Angles

KEY Concept

Angles can be described as complementary or supplementary, according to their sums.

Complementary Angles

∠ABC and ∠DBA are complementary.			
measure of the angles	40°	50°	90°
equation	40°	+ 50°	= 90°

Complementary angles with a common ray form a right angle.

Supplementary Angles

∠XYZ and ∠RYX are supplementary.			
measure of the angles	115°	65°	180°
equation	115°	+ 65°	= 180°

Supplementary angles with a common ray form a straight angle.

Angles in Figures

The sum of the measures of the angles of a triangle is 180°.		95° + 50° + 35° = 180°
The sum of the measures of the angles of a quadrilateral is 360°.		180° + 180° = 360°

Copyright © Glencoe/McGraw-Hill, a division of The McGraw-Hill Companies, Inc.

VOCABULARY

complementary angles
two angles are complementary if the sum of their measures is 90°

right angle
an angle that measures 90°

straight angle
an angle that measures exactly 180°

supplementary angles
two angles are supplementary if the sum of their measures is 180°

A protractor can be used to measure and sketch angles.

Example 1

Find the measure of the missing angle.

1. The measure of ∠FHJ equals 180°.

2. ∠FHG and ∠GHJ are supplementary angles.

3. Find $m\angle GHJ$.

$$m\angle FHG + m\angle GHJ = 180$$
$$100° + ?° = 180$$
$$?° = 80$$

4. $m\angle GHJ = 80°$

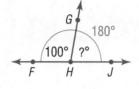

YOUR TURN!

Find the measure of the missing angle.

1. The measure of ∠ACD equals _____.

2. ∠ACB and ∠BCD are complementary angles.

3. Find $m\angle ACB$.

$$m\angle BCD + m\angle ACB = \underline{\hspace{1cm}}$$
$$\underline{\hspace{1cm}} + ?° = \underline{\hspace{1cm}}$$
$$?° = \underline{\hspace{1cm}}$$

4. $m\angle ACB = \underline{\hspace{1cm}}$

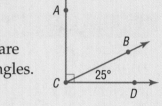

Example 2

Find the measure of the missing angle.

1. Find the sum of the measures of the known angles.

$$m\angle G + m\angle H$$
$$= 75° + 50°$$
$$= 125°$$

2. The sum of the measures of the angles of a triangle is 180°. Subtract the sum of the known measures from 180°.

$$180° - 125° = 55°$$

The measure of the missing angle is 55°.

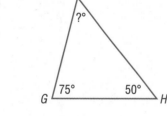

YOUR TURN!

Find the measure of the missing angle.

1. Find the sum of the measures of known angles.

$$m\angle Q + m\angle R + m\angle T$$
$$= \underline{\hspace{1cm}} + \underline{\hspace{1cm}} + \underline{\hspace{1cm}}$$
$$= \underline{\hspace{1cm}}$$

2. The sum of the measures of the angles of a quadrilateral is _____. Subtract the sum of the known measures from _____.

$$\underline{\hspace{1cm}} - \underline{\hspace{1cm}} = \underline{\hspace{1cm}}$$

The measure of the missing angle is _____.

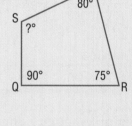

Copyright © Glencoe/McGraw-Hill, a division of The McGraw-Hill Companies, Inc.

Example 3

Sketch supplementary angles when one angle's measure is 45°.

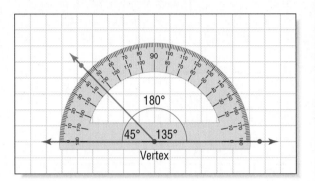

1. The sum of supplementary angles is 180°. Sketch a straight angle.

2. Indicate the vertex of the straight angle.

3. Use one ray of the straight angle and create an angle with a measure of 45°.

4. Measure the remaining angle. The measure of the angle is 135°.

5. Check your answer.
   180° − 45° = 135°

Copyright © Glencoe/McGraw-Hill, a division of The McGraw-Hill Companies, Inc.

YOUR TURN!

Sketch complementary angles when one angle's measure is 70°.

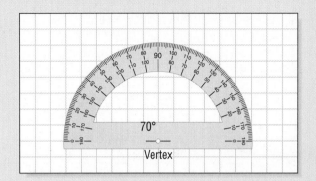

1. The sum of supplementary angles is _____. Sketch a _____ angle.

2. Indicate the vertex of the _____ angle.

3. Use one ray of the right angle. Create an angle with a measure of _____.

4. Measure the remaining angle. The measure of the angle is _____.

5. Check your answer.
   _____ − _____ = _____

Who is Correct?

∠MNO and ∠PNQ are supplementary angles. The measure of ∠MNO is 67°. What is the measure of ∠PNQ?

Diego
90° − 67° = 23°

Tionne
180° − 67° = 123°

Malia
180° − 67° = 113°

Circle correct answer(s). Cross out incorrect answer(s).

Find the measure of the missing angle.

1

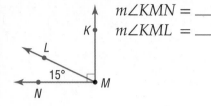

$m\angle KMN =$ _____
$m\angle KML =$ _____

2

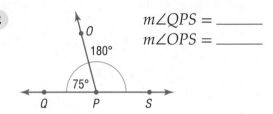

$m\angle QPS =$ _____
$m\angle OPS =$ _____

Step (by) **Step Practice**

3 **Find the measure of the missing angle.**

Step 1 Find the sum of the measures of known angles.

$m\angle D + m\angle E =$ _____ + _____ = _____

Step 2 The sum of the measures of the angles of a triangle is _____. Subtract the sum of measures of the known angles from _____.

_____ − _____ = _____

The measure of the missing angle is _____.

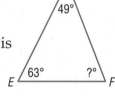

Find the measure of each missing angle.

4

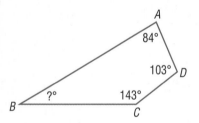

$m\angle A +\ \ m\angle C +\ \ m\angle D =$

_____ + _____ + _____ = _____

_____ − _____ = _____

The measure of the missing angle is _____.

5

$m\angle K =$ _____

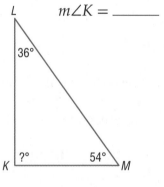

Copyright © Glencoe/McGraw-Hill, a division of The McGraw-Hill Companies, Inc.

Sketch each type of angle given.

6 Sketch supplementary angles when one angle's measure is 85°.

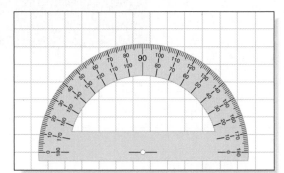

7 Sketch complementary angles when one angle's measure is 41°.

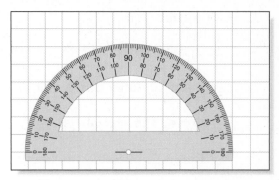

Step *by* Step *Problem-Solving Practice*

Solve.

8 **ART** Samir drew a version of a pine tree in the shape of a triangle. It had a 36° angle and a 70° angle. What was the measure of the third angle of Samir's pine tree?

Understand	Read the problem. Write what you know. The measures of the known angles are _____ and _____.
Plan	Pick a strategy. One strategy is to solve a simpler problem.
Solve	Subtract the sum of the known angles from 180°. _____ + _____ = _____ 180° − _____ = _____ The measure of the third angle is _____.
Check	The sum of the three angle measures must equal _____. 36° + 70° + _____ = _____

Problem-Solving Strategies
☐ Draw a diagram.
☐ Look for a pattern.
☐ Guess and check.
☐ Act it out.
☑ Solve a simpler problem.

Copyright © Glencoe/McGraw-Hill, a division of The McGraw-Hill Companies, Inc.

GO ON

9 AGRICULTURE A cow pasture is shaped like a parallelogram. Three of the angles measure 68°, 68°, and 112°. What is the measure of the fourth angle?

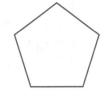

Check off each step.

_____ **Understand: I underlined key words.**

_____ **Plan: To solve this problem, I will** _____.

_____ **Solve: The answer is** _____.

_____ **Check: I checked my answer by** _____.

10 SPORTS Hughes Hardware engraves trophies. The trophies have sides in the shape of isosceles triangles. The two base angles each measure 73°. What is the measure of the top angle?

_____

11 Reflect Explain how you can find the sum of the angle measures in a pentagon.

_____

_____

_____

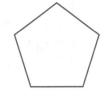

▶ Skills, Concepts, and Problem Solving

Find the measure of the missing angle.

12

$m\angle KLN =$ _____
$m\angle MLN =$ _____

13

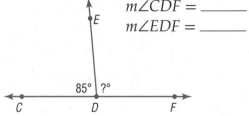

$m\angle CDF =$ _____
$m\angle EDF =$ _____

Find the measure of each missing angle.

14

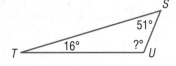

$m\angle U =$ _____

15

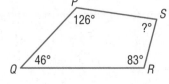

$m\angle S =$ _____

Copyright © Glencoe/McGraw-Hill, a division of The McGraw-Hill Companies, Inc.

Find the measure of the missing angle.

16
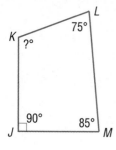

$m\angle K = \underline{\hspace{1.5cm}}$

17
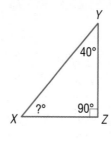

$m\angle X = \underline{\hspace{1.5cm}}$

Sketch each type of angle given.

18 Sketch complementary angles when one angle's measure is 58°.

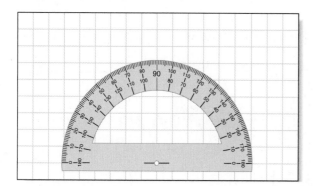

19 Sketch supplementary angles when one angle's measure is 52°.

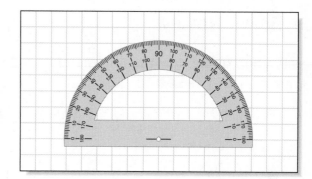

Solve.

20 **GARDENING**   The butterfly garden at Sweet Valley Academy is placed in the corner of the courtyard. It is in the shape of a triangle. The butterfly garden has angles that measure 62° and 47°. What is the measure of the third angle? _____

21 **SCULPTURE**   The geometric sculpture that sits at the entrance of Ugo Park has a triangle for its base. It has angles that measure 38° and 104°. What is the measure of the third angle? _____

22 **PETS**   The play area at Pooch Paradise is in the shape of a quadrilateral. Three angles of the area measure 25°, 112°, and 82°. What is the measure of the fourth angle? _____

23 **CROSS-COUNTRY**   A cross-country team ran on a path near the school that was in the shape of a quadrilateral. Three angles of the path measured 42°, 85°, and 102°. What is the measure of the fourth angle? _____

Copyright © Glencoe/McGraw-Hill, a division of The McGraw-Hill Companies, Inc.

Vocabulary Check **Write the vocabulary word that completes each sentence.**

24 _____ angles are two angles that have measures with a sum of 90°.

25 _____ angles are two angles that have measures with a sum of 180°.

26 **Writing in Math** Explain how to find the missing angle of a quadrilateral when the measures of three angles are given.

_____

_____

_____

▶ Spiral Review

Classify each triangle by the lengths of its sides and the measures of its angles. (Lesson 4-2, p. 145)

27
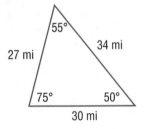

_____ triangle

28
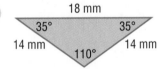

_____ triangle

Sketch a figure with the description given. (Lesson 4-2, p. 145)

29 acute, isosceles triangle

30 right, scalene triangle

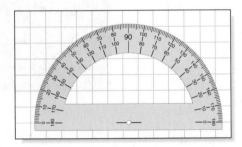

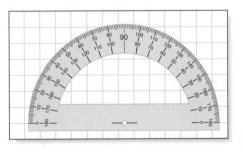

Copyright © Glencoe/McGraw-Hill, a division of The McGraw-Hill Companies, Inc.

Transversals

KEY Concept

In the diagram below, the ray creates supplementary angles *A* and *B*.

A similar relationship occurs when two lines intersect. When line ℓ intersects line *m* four angles are created.

∠*A* and ∠*D* are **vertical angles**. They are opposite angles that were formed when the two lines intersected. ∠*B* and ∠*C* are also vertical angles. Each pair of vertical angles are congruent.

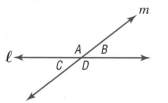

$$\angle A \cong \angle D$$

$$\angle B \cong \angle C$$

Other angle relationships occur when a transversal passes through two parallel lines. The table below lists these relationships.

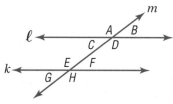

Angle Relationships	
alternate interior	∠*C* and ∠*F* ∠*D* and ∠*E*
alternate exterior	∠*A* and ∠*H* ∠*B* and ∠*G*
corresponding	∠*A* and ∠*E* ∠*B* and ∠*F* ∠*C* and ∠*G* ∠*D* and ∠*H*
vertical	∠*A* and ∠*D* ∠*B* and ∠*C* ∠*E* and ∠*H* ∠*F* and ∠*G*

VOCABULARY

alternate exterior angles
exterior angles that lie on opposite sides of the transversal

alternate interior angles
interior angles that lie on opposite sides of the transversal

congruent angles
angles with the same measure

corresponding angles
angles that have the same position on two different parallel lines cut by a transversal

parallel lines
lines that do not intersect

transversal
a line that intersects two or more lines to form eight angles

vertical angles
nonadjacent angles formed by a pair of lines that intersect

GO ON

Copyright © Glencoe/McGraw-Hill, a division of The McGraw-Hill Companies, Inc.

Copyright © Glencoe/McGraw-Hill, a division of The McGraw-Hill Companies, Inc.

Example 1

Identify the measure of ∠K if the measure of ∠H is 100°.

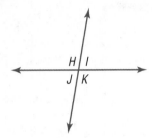

1. Identify two pairs of vertical angles.

   ∠H and ∠K
   ∠I and ∠J

2. To find the measure of the missing angle, find the measure of the opposite angle.

   ∠H ≅ ∠K

3. The measure of ∠K is 100°.

YOUR TURN!

Identify the measure of ∠N if the measure of ∠M is 15°.

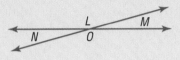

1. Identify two pairs of vertical angles.

   _____ and _____

   _____ and _____

2. To find the measure of the missing angle, find the measure of the opposite angle.

   ∠_____ ≅ ∠N

3. The measure of ∠N is _____.

Example 2

Name the alternate interior angles.

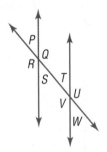

1. Name the angles that are between the two parallel lines.

   ∠Q, ∠T, ∠S, and ∠V

2. Identify two sets of alternate angles.

   ∠Q and ∠V
   ∠S and ∠T

YOUR TURN!

Name the alternate exterior angles.

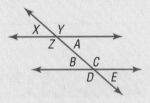

1. Name the angles that are on the outside of the two parallel lines.

   _____, _____, _____ and _____

2. Identify two sets of alternate angles.

   _____ and _____

   _____ and _____

Example 3

Find the value of x if the given angle equals 125°.

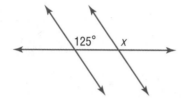

1. Identify the relationship between the given angle and angle x.

   The given angle and $\angle x$ are corresponding angles.

2. The measures of these two angles are equal.

3. The value of x is 125°.

YOUR TURN!

Find the value of x if the given angle equals 20°.

1. Identify the relationship between the given angle and angle x.

   The given angle and $\angle x$ are _____ _____.

2. The measures of these two angles _____.

3. The value of x is _____.

Who is Correct?

Name the alternate exterior angles.

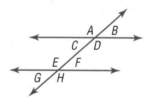

Jermaine
∠A and ∠H
∠B and ∠G

Allison
∠C and ∠F
∠D and ∠E

Dasan
∠A and ∠C
∠E and ∠G

Circle correct answer(s). Cross out incorrect answer(s).

Guided Practice

Identify the measure of each indicated angle.

1.
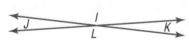

   $m\angle J = 10°$, so $m\angle K =$ _____.

2.

   $m\angle M = 117°$, so $m\angle P =$ _____.

GO ON

Copyright © Glencoe/McGraw-Hill, a division of The McGraw-Hill Companies, Inc.

Step (by) Step Practice

3 Identify the measure of ∠*T* if ∠*R* equals 85°.

Step 1 Identify two pairs of supplementary angles.

∠_____ and ∠_____

∠_____ and ∠_____

Step 2 To find the measure of the missing angle, use the measures of the supplementary angles.

$m\angle$_____ + $m\angle$_____ = _____

_____ – _____ = _____

Step 3 The measure of ∠_____ is _____.

Identify the measure of each angle indicated.

4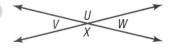

$m\angle V = 26°$, so $m\angle W =$ _____.

5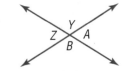

$m\angle Z = 63°$, so $m\angle B =$ _____.

Name the alternate interior angles.

6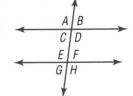

_____ and _____ _____ and _____

7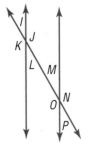

_____ and _____ _____ and _____

Name the alternate exterior angles.

8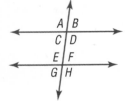

_____ and _____ _____ and _____

9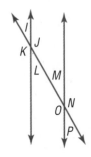

_____ and _____ _____ and _____

Copyright © Glencoe/McGraw-Hill, a division of The McGraw-Hill Companies, Inc.

Step by Step Problem-Solving Practice

Problem-Solving Strategies
☐ Draw a diagram.
☐ Use logical reasoning.
☐ Solve a simpler problem.
☐ Work backward.
☑ Look for a pattern.

Solve.

10 STAIRS   Mr. Hataro is building a staircase in a new home. The rails are parallel. He checks the rails by measuring the angles. If the measure of ∠Q is 85°, what is the value of x?

Understand	Read the problem. Write what you know.
	The rails are _____.
	$m\angle Q = $ _____
Plan	Pick a strategy. One strategy is to look for a pattern.
Solve	Identify the relationship between ∠Q and the value of x.
	The given angle and ∠x are _____.
	The measures of these two angles _____.
	The value of x is _____.
Check	Use a different angle relationship to solve the problem.

11 RAILROADS   The wooden ties of a railroad track are parallel. What is the value of x if the m∠R is 88°? Check off each step.

_____ Understand: I underlined key words.

_____ Plan: To solve this problem, I will _____.

_____ Solve: The answer is _____.

_____ Check: I checked my answer by _____.

12 QUILTS   Mrs. Hawkins is making a quilt. The lines shown on the quilt are parallel. What is the value of x if the measure of ∠S is 30°?

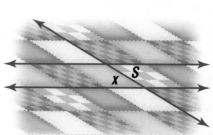

_____

13 Reflect   If you know one of the angle measures in the quilt sample, can you find the measures of the remaining angles without using a protractor?

_____

_____

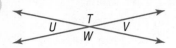

Skills, Concepts, and Problem Solving

Identify the measure of each missing angle.

14

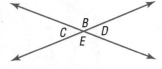

$m\angle U = 23°$, so $m\angle V =$ _____.

15

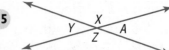

$m\angle X = 146°$, so $m\angle Z =$ _____.

16

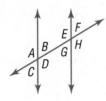

$m\angle D = 42°$, so $m\angle E =$ _____.

17

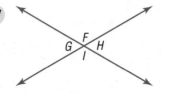

$m\angle I = 121°$, so $m\angle G =$ _____.

Name the alternate interior angles.

18

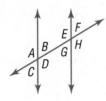

_____ and _____   _____ and _____

19

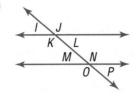

_____ and _____   _____ and _____

Name the alternate exterior angles.

20

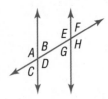

_____ and _____   _____ and _____

21

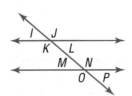

_____ and _____   _____ and _____

Solve.

22 **MAPS** First Street and Second Street run parallel to one another. Main Street intersects both of them. What is the value of x if the measure of $\angle Q$ is 84°?

_____

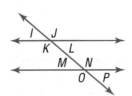

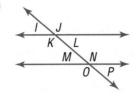

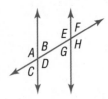

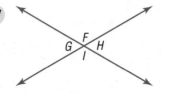

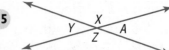

Copyright © Glencoe/McGraw-Hill, a division of The McGraw-Hill Companies, Inc.

23 **BRIDGES** The top beam and the floor of the bridge are parallel. What is the value of x if the measure of ∠S is 45°?

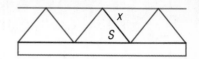

_____

Vocabulary Check **Write the vocabulary word that completes each sentence.**

24 _____ angles are angles with the same measure.

25 A(n) _____ is a line that intersects two or more lines to form eight angles.

26 **Writing in Math** If two parallel lines are cut by a transversal, what relationship exists between alternate exterior angles? Explain.

_____

 Spiral Review

Sketch supplementary angles when one angle's measure is 65°.

(Lesson 4-3, p. 153)

27

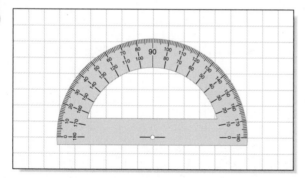

Find the measure of each missing angle. (Lesson 4-2, p. 145)

28

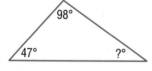

_____

29

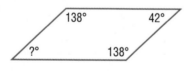

_____

Solve. (Lesson 4-2, p. 145)

30 **TENTS** The doorway to Claudio's new tent is in the shape of a quadrilateral. Three angles of the doorway measure 123°, 123°, and 57°. What is the measure of the missing angle?

_____

Copyright © Glencoe/McGraw-Hill, a division of The McGraw-Hill Companies, Inc.

Draw each type of angle given.

1 Draw supplementary angles when one angle's measure is 120°.

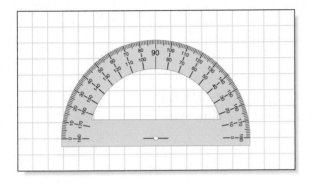

2 Draw complementary angles when one angle's measure is 45°.

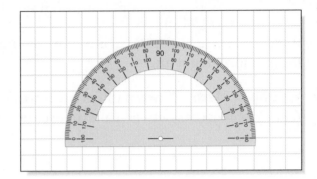

Find the measure of each missing angle.

3 $m\angle H =$ _____

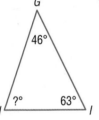

4 $m\angle L =$ _____

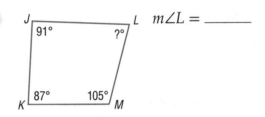

Identify the measure of each angle indicated.

5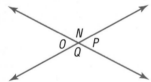

$m\angle N = 126°$, so $m\angle O =$ _____.

6

$m\angle R = 149°$, so $m\angle T =$ _____.

Solve.

7 **REPAIRS** A school bus was placed on a lift so that repair workers could see under the bus. The vertical lift bars are parallel to one another and the horizontal bars are parallel to one another. If the measure of $\angle B$ is 113°, what is the value of x?

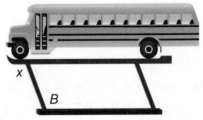

Copyright © Glencoe/McGraw-Hill, a division of The McGraw-Hill Companies, Inc.

STOP

Vocabulary and Concept Check

acute angle, *p. 145*

angle, *p. 138*

complementary angles, *p. 153*

congruent, *p. 145*

degree, *p. 138*

obtuse angle, *p. 145*

parallel lines, *p. 161*

protractor, *p. 138*

right angle, *p. 145*

side, *p. 145*

straight angle, *p. 153*

supplementary angles, *p. 153*

triangle, *p. 145*

vertex, *p. 138*

Write the vocabulary word that completes each sentence.

1 A(n) _____ is an instrument marked in degrees, used for measuring or drawing angles.

2 A(n) _____ is a unit for measuring angles.

3 Two angles are _____ if the sum of their measures is 180°.

4 Two rays with a common endpoint form an _____.

5 A(n) _____ angle measures greater than 90°, but less than 180°.

Label each diagram below. Write the correct vocabulary term in each blank.

6 _____

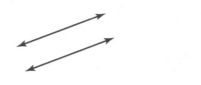

7 _____

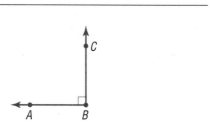

8 _____

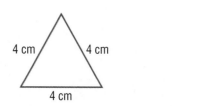

9 _____

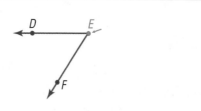

10 _____

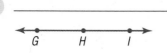

11 _____

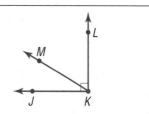

Copyright © Glencoe/McGraw-Hill, a division of The McGraw-Hill Companies, Inc.

Lesson Review

4-1 Angles (pp. 138–144)

12 Draw ∠MNO that measures 135°.

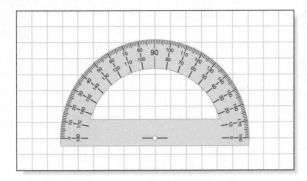

13 Draw ∠PQR that measures 40°.

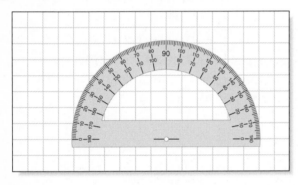

Measure and identify the angle.

14

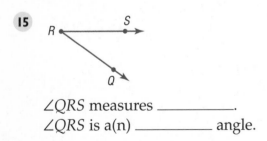

∠NOP measures _____.

∠NOP is a(n) _____ angle.

15

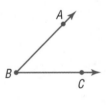

∠QRS measures _____.

∠QRS is a(n) _____ angle.

Example 1

Draw ∠JKL that measures 55°.

1. Draw $\overrightarrow{KL}$.

2. Place point K as the vertex.

3. Use the inner scale and draw point J at 55°.

4. Draw $\overrightarrow{KJ}$.

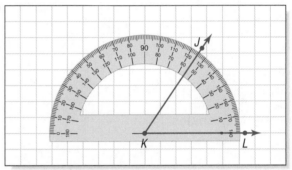

Example 2

Measure and identify the angle.

1. Place the center of the protractor at point B. Line up $\overrightarrow{BC}$ with the line that extends from 0° to 180° on the protractor.

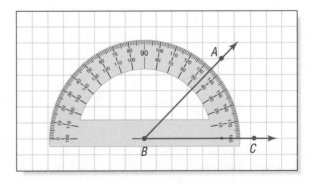

2. Look at point A. Read the measure of the angle where $\overrightarrow{BA}$ passes through the inner scale. ∠ABC measures 45°.

Copyright © Glencoe/McGraw-Hill, a division of The McGraw-Hill Companies, Inc.

4-2 Triangles (pp. 145–151)

Classify each triangle by the lengths of its sides.

16.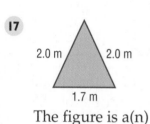

3 ft
90°
4 ft
5 ft

The figure is a(n)

_____.

17.

2.0 m   2.0 m

1.7 m

The figure is a(n)

_____.

Classify each triangle by the measure of its angles.

18.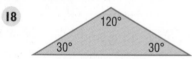

120°
30°   30°

The figure is a(n)

_____.

19.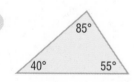

85°
40°   55°

The figure is a(n)

_____.

Copyright © Glencoe/McGraw-Hill, a division of The McGraw-Hill Companies, Inc.

Example 3

Classify the triangle by the lengths of its sides.

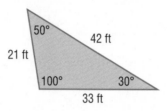

50°
42 ft
21 ft
100°   30°
33 ft

1. The side lengths are 21 feet, 42 feet, and 33 feet.

2. None of the sides of the triangle are equal in length. The figure is a scalene triangle.

Example 4

Classify triangle _MNO_ by the measure of its angles.

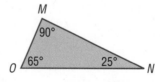

M
90°
O   65°   25°   N

1. $m\angle O = 65°$ $\angle O$ is an acute angle.
2. $m\angle M = 90°$ $\angle M$ is a right angle.
3. $m\angle N = 25°$ $\angle N$ is an acute angle.
4. The figure is a right triangle.

4-3 Add Angles (pp. 153–160)

Find the measure of each missing angle.

20

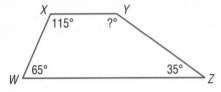

The measure of the missing angle

is _____.

21

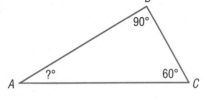

The measure of the missing angle

is _____.

Find the measure of the missing angle.

22

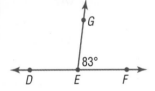

$m\angle DEF = $ _____

$m\angle GED = $ _____

23

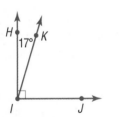

$m\angle HIJ = $ _____

$m\angle KIJ = $ _____

Example 5

What is the measure of the missing angle?

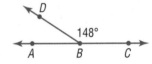

1. Find the sum of the measures of known angles.

   $m\angle Q + m\angle R + m\angle S = 300°$

2. The sum of the measures of the angles of a quadrilateral is 360°. Subtract the sum of the known measures from 360°.

   $360° - 300° = 60°$

Example 6

Find the measure of the missing angle.

1. The measure of $\angle ABC$ equals $180°$.

2. Find $m\angle ABD$.

   $m\angle ABD + m\angle DBC = 180°$

   $?° \quad + \quad 148° = 180°$

   $?° \quad = 32°$

3. $m\angle ABD = 32°$

Copyright © Glencoe/McGraw-Hill, a division of The McGraw-Hill Companies, Inc.

4-4 Transversals (pp. 161–167)

Identify the measure of each indicated angle.

24 $m\angle E = 86°$, so $m\angle H =$ _____.

25 $m\angle E = 86°$, so $m\angle F =$ _____.

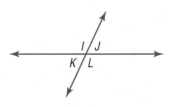

26 $m\angle K = 64°$, so $m\angle J =$ _____.

27 $m\angle K = 64°$, so $m\angle L =$ _____.

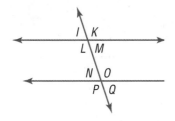

28 Name the alternate exterior angles.

_____ and _____

_____ and _____

29 Name the alternate interior angles.

_____ and _____

_____ and _____

Example 7

Identify the measure of ∠D if the measure of ∠A is 97°.

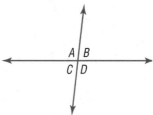

1. ∠A and ∠D are vertical angles.
   ∠B and ∠C are vertical angles.

2. Find the measure of the opposite angle.

   ∠A ≅ ∠D

3. The measure of ∠D is 97°.

Example 8

Name the alternate interior angles.

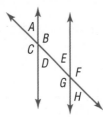

1. Name the angles that are between the two parallel lines.

   ∠B, ∠E, ∠D, and ∠G

2. Identify two sets of alternate angles.

   ∠B and ∠G
   ∠D and ∠E

Copyright © Glencoe/McGraw-Hill, a division of The McGraw-Hill Companies, Inc.

Measure and identify each angle.

1
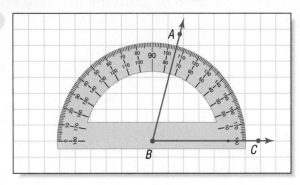

∠ABC measures _____.

∠ABC is a(n) _____.

2
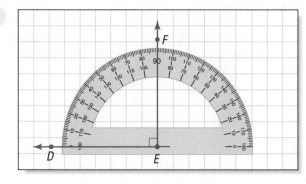

∠DEF measures _____.

∠DEF is a(n) _____.

Draw each type of angle given.

3 complementary angles
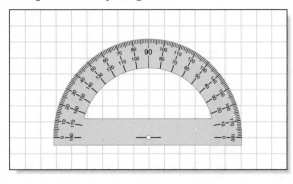

4 supplementary angles
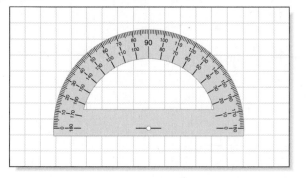

Find the measure of the missing angle.

5
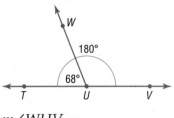

m∠WUV = _____.

6

m∠XYA = _____.

7
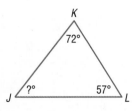

m∠J = _____.

8
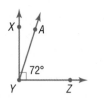

m∠P = _____.

Copyright © Glencoe/McGraw-Hill, a division of The McGraw-Hill Companies, Inc.

Identify the measure of each missing angle.

9

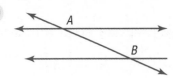

$m\angle U = 32°$, so $m\angle V =$ _____.

10

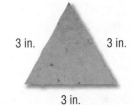

$m\angle A = 157°$, so $m\angle B =$ _____.

Solve.

11 CONSTRUCTION Angela had to fit two pieces of crown molding together at a 125° angle. What type of angle did the molding form?

_____

12 COOKING Jada made some fried corn tortillas like the one pictured at the right. What type of triangle do Jada's tortillas represent?

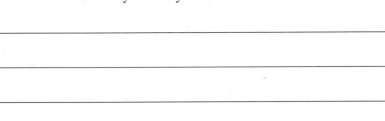

_____

13 LABELS Ms. Saracino is designing a label for a lotion bottle. The label has four sides. Three angles have measures 72°, 72°, and 90°. What is the measure of the missing angle?

_____

14 DOORS The door of Emilio's room makes a right angle with the wall. His door will open at a 73° angle. What is the angle between his door and the wall?

_____

Correct the mistakes.

15 Saba says that $\angle P$ and $\angle Q$ are congruent angles. Is she correct? Why or why not?

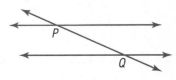

_____

_____

_____

_____

STOP

Copyright © Glencoe/McGraw-Hill, a division of The McGraw-Hill Companies, Inc.

Chapter 4 · Test Practice

Choose the best answer and fill in the corresponding circle on the sheet at right.

1 Which best describes these lines?

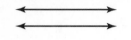

    A perpendicular   **C** bisecting

    B intersecting   **D** parallel

2 What is the approximate measure of this angle?

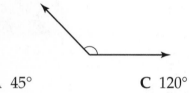

    A 45°   **C** 120°

    B 90°   **D** 175°

3 What is the value of x in the figure?

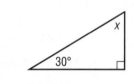

    A 30°   **C** 60°

    B 45°   **D** 150°

4 What is the measure of $\angle x$?

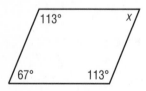

    A 67°   **C** 180°

    B 113°   **D** 360°

5 What type of angle is $\angle ABC$?

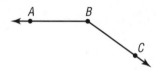

    A right   **C** congruent

    B obtuse   **D** acute

6 What is the approximate measure of angle $\angle QRS$?

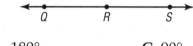

    A 180°   **C** 90°

    B 165°   **D** 45°

7 What is the sum of the measures of the angles of a triangle?

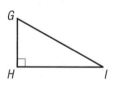

    A 45°   **C** 180°

    B 90°   **D** 360°

Copyright © Glencoe/McGraw-Hill, a division of The McGraw-Hill Companies,

8 What is the value of *x*?

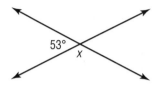

A 37°

C 127°

B 53°

D 180°

---

9 What term describes angles *EFG* and *GFH*?

A complementary

C obtuse

B congruent

D supplementary

---

10 What term does NOT describe the figure below?

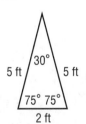

A triangle

C isosceles

B obtuse

D acute

11 What is the measure of ∠*A* and ∠*B*?

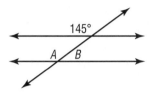

A 145°, 35°

C 55°, 35°

B 145°, 65°

D 140°, 40°

ANSWER SHEET

Directions: Fill in the circle of each correct answer.

1 Ⓐ Ⓑ Ⓒ Ⓓ

2 Ⓐ Ⓑ Ⓒ Ⓓ

3 Ⓐ Ⓑ Ⓒ Ⓓ

4 Ⓐ Ⓑ Ⓒ Ⓓ

5 Ⓐ Ⓑ Ⓒ Ⓓ

6 Ⓐ Ⓑ Ⓒ Ⓓ

7 Ⓐ Ⓑ Ⓒ Ⓓ

8 Ⓐ Ⓑ Ⓒ Ⓓ

9 Ⓐ Ⓑ Ⓒ Ⓓ

10 Ⓐ Ⓑ Ⓒ Ⓓ

11 Ⓐ Ⓑ Ⓒ Ⓓ

Success Strategy

If you do not know the answer to a question, go on to the next question. Come back to the problem, if you have time. You might find another question later in the test that will help you figure out the skipped problem.

STOP

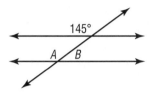

Copyright © Glencoe/McGraw-Hill, a division of The McGraw-Hill Companies, Inc.

Ratios, Rates, and Similarity

Have you ever checked your heart rate?

Your heart rate describes the number of heartbeats per minute. Doctors can gather important information by comparing an individual's heart rate to average heart rates. Athletes usually have a lower heart rate than non-athletes.

Copyright © Glencoe/McGraw-Hill, a division of The McGraw-Hill Companies, Inc.

STEP 1 Quiz

Math Online ▷ Are you ready for Chapter 5? Take the Online Readiness Quiz at *glencoe.com* to find out.

STEP 2 Preview

Get ready for Chapter 5. Review these skills and compare them with what you'll learn in this chapter.

What You Know	What You Will Learn

What You Know

You know how to write fractions to represent parts of a group.

$\frac{4}{5}$ of the sections are blue.

TRY IT!

What fraction is blue?

1    **2**

_____   _____

You know how to simplify fractions.

Example: $\dfrac{171 \div 3}{3 \div 3} = \dfrac{57}{1} = 57$

TRY IT!

Simplify each fraction.

3 $\dfrac{360}{12} =$ _____

4 $\dfrac{195}{13} =$ _____

You know how to multiply.

Example: $40 \times 30 = 1{,}200$

TRY IT!

5 $2{,}000 \times 6 =$ _____

6 $16 \times 20 =$ _____

7 $5{,}280 \times 5 =$ _____

What You Will Learn

Lesson 5-1

Ratios are a way to compare numbers. A common way to write a ratio is as a fraction in simplest form.

There are 4 blue sections for every 5 sections.

The ratio of blue sections to the whole is $\frac{4}{5}$.

The ratio of blue sections to yellow sections is $\frac{4}{1}$.

Lesson 5-2

A **rate** is a ratio that compares different units. When a rate has a denominator of 1, it is a **unit rate**.

A snack pack of 4 crackers has 64 Calories.

$$\frac{64 \text{ Calories}}{4 \text{ crackers}} = \frac{16 \text{ Calories}}{1 \text{ cracker}}$$

Each cracker has 16 Calories.

Lesson 5-3

Proportions are equivalent ratios.

$$\frac{7}{8} = \frac{28}{32}$$

To solve a proportion, **cross multiply**.

$$\frac{12}{1} = \frac{n}{6}$$

$1 \cdot n = 6 \cdot 12 \qquad n = 72$

Copyright © Glencoe/McGraw-Hill, a division of The McGraw-Hill Companies, Inc.

Ratios

KEY Concept

Ratios are a way to compare numbers. A **ratio** is a comparison of two quantities by division. Ratios can compare a part to a part, a part to a whole, or a whole to a part.

Look at the following pattern.

The pattern shows 6 triangles out of 10 figures.

The ratio of triangles to figures is $\frac{6}{10}$.

Ratios are often written in **simplest form**. To find the simplest form you need the **greatest common factor**. The greatest common factor of 6 and 10 is 2.

$$\frac{6}{10} = \frac{6}{10} \div \frac{2}{2} = \frac{3}{5}$$

Other ways to write the ratio of triangles to figures are:

$$3 \text{ to } 5 \qquad 3 \text{ out of } 5 \qquad 3{:}5$$

VOCABULARY

greatest common factor (GCF)
the greatest of the common factors of two or more numbers; the GCF of 24 and 30 is 6

ratio
a comparison of two numbers by division; the ratio of 2 to 3 can be stated as 2 out of 3, 2 to 3, 2:3, or $\frac{2}{3}$

simplest form
the form of a fraction when the GCF of the numerator and the denominator is 1

Example 1

Write the ratio that compares the number of dimes to the number of pennies. Explain the meaning of the ratio.

1. Write the ratio with the number of dimes in the numerator and the number of pennies in the denominator.

   $\frac{3}{5}$ ← dimes
   ← pennies

2. The only common factor of 3 and 5 is 1. The ratio is in simplest form.

3. The ratio of the number of dimes to the number of pennies is written as $\frac{3}{5}$, 3 to 5, or 3:5.

4. The ratio means *for every 3 dimes, there are 5 pennies.*

Copyright © Glencoe/McGraw-Hill, a division of The McGraw-Hill Companies, Inc.

Copyright © Glencoe/McGraw-Hill, a division of The McGraw-Hill Companies, Inc.

YOUR TURN!

Write the ratio that compares the number of pennies to the total number of coins. Explain the meaning of the ratio.

1. Write the ratio.

$$\frac{\boxed{}}{\boxed{}}$$ ← pennies
← total coins

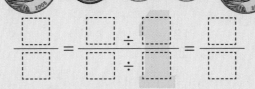

2. The numerator and denominator have a common factor of _____.
Write the fraction in simplest form.

$$\frac{\boxed{}}{\boxed{}} = \frac{\boxed{} \div \boxed{}}{\boxed{} \div \boxed{}} = \frac{\boxed{}}{\boxed{}}$$

3. Write the ratio of the number of pennies to the number of coins.

_____

4. What does the ratio mean? _____

_____

Example 2

Write the ratio as a fraction in simplest form.

4 boys out of 12 children

1. Write the ratio with the number of boys in the numerator and the total number of children in the denominator.

$$\frac{4}{12}$$

2. The greatest common factor of 4 and 12 is 4. Divide by 4 to write the fraction in simplest form.

$$\frac{4}{12} = \frac{4 \div 4}{12 \div 4} = \frac{1}{3}$$

YOUR TURN!

Write the ratio as a fraction in simplest form.

12 children to 18 adults

1. Write the ratio.

$$\frac{\boxed{}}{\boxed{}}$$

2. The greatest common factor of 12 and 18 is _____. Write the fraction in simplest form.

$$\frac{\boxed{}}{\boxed{}} = \frac{\boxed{} \div \boxed{}}{\boxed{} \div \boxed{}} = \frac{\boxed{}}{\boxed{}}$$

GO ON

Example 3

Write the ratio of the length of the base to the height of the triangle as a fraction in simplest form.

height = 3 cm
base = 18 cm

1. Write the ratio as a fraction with the **base** over the **height**.

$$\frac{18}{3}$$

2. The greatest common factor of 3 and 18 is 3. Divide by 3 to write the fraction in simplest form.

$$\frac{18}{3} = \frac{18 \div 3}{3 \div 3} = \frac{6}{1}$$

YOUR TURN!

Write the ratio of the length of the base to the height of the triangle as a fraction in simplest form.

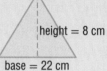

height = 8 cm
base = 22 cm

1. Write the ratio.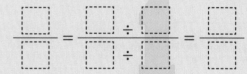

2. The greatest common factor of 22 and 8 is _____. Write the fraction in simplest form.

$$\frac{\Box}{\Box} = \frac{\Box \div \Box}{\Box \div \Box} = \frac{\Box}{\Box}$$

Who is Correct?

Write the ratio as a fraction in simplest form.

96 boys to 8 girls

Pamela
$$\frac{96}{8} = \frac{24}{4}$$

Mateo
$$\frac{96}{8} = \frac{12}{2}$$

Rodrigo
$$\frac{96}{8} = \frac{12}{1}$$

Circle correct answer(s). Cross out incorrect answer(s).

▶ Guided Practice

Use the diagram to write each ratio as a fraction in simplest form.

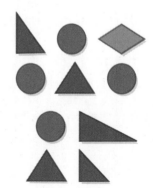

1 The number of triangles to the number of circles is _____.

2 The number of triangles to the total number of figures is _____.

3 The number of circles to the total number of figures is _____.

Copyright © Glencoe/McGraw-Hill, a division of The McGraw-Hill Companies, Inc.

4   **ANIMALS**   A petting zoo has 8 rabbits, 3 goats, 6 ducks, and 7 sheep. Write the ratio of each type of animal to the total number of animals in the zoo. Write each as a fraction in simplest form.

Step 1   The total number of animals is _____.
This will be the _____ in each fraction.

Step 2   Write a ratio for the number of rabbits to the total number of animals. _____

What is the common factor of the numerator and denominator? _____

Write the fraction in simplest form. $\dfrac{\Box}{\Box} = \dfrac{\Box \div \Box}{\Box \div \Box} = \dfrac{\Box}{\Box}$

Step 3   Write a ratio for the number of goats to the total number of animals. _____

What is the common factor of the numerator and denominator? _____

Write the fraction in simplest form. $\dfrac{\Box}{\Box} = \dfrac{\Box \div \Box}{\Box \div \Box} = \dfrac{\Box}{\Box}$

Step 4   Write a ratio for the number of ducks to the total number of animals. _____

What is the common factor of the numerator and denominator? _____

Write the fraction in simplest form. $\dfrac{\Box \div \Box}{\Box \div \Box} = \dfrac{\Box}{\Box}$

Step 5   Write a ratio for the number of sheep to the total number of animals. _____

What is the common factor of the numerator and denominator? _____

Write the fraction in simplest form. _____

Copyright © Glencoe/McGraw-Hill, a division of The McGraw-Hill Companies, Inc.

Write each ratio as a fraction in simplest form.

5   **FLOWERS**   In a vase of flowers, there are 6 tulips and 9 roses. Write the ratio of tulips to roses.

tulips →
roses → $\dfrac{\Box}{\Box} = \dfrac{\Box \div \Box}{\Box \div \Box} = \dfrac{\Box}{\Box}$

6 **COMPUTERS** In a classroom, there are 32 students and 6 computers. Write the ratio of students to computers.

_____

7 **BASKETS** In a shipment of 18 baskets, there are 16 that are not white. Write the ratio of white baskets to nonwhite baskets.

_____

8 **SPORTS** In a sports equipment closet, there are 6 baseballs, 4 basketballs, 4 footballs, and 7 soccer balls. Write the ratio of soccer balls to the total number of balls.

_____

Step by Step Problem-Solving Practice

Solve.

9 **MUSIC** Freeman has 30 CDs, and his sister Iesha has 18 CDs. If they both buy two more CDs, what will be the ratio of Freeman's CDs to Iesha's CDs?

Understand	Read the problem. Write what you know.
	Freeman has _____ CDs.
	Iesha has _____ CDs.
	If they both buy two more, Freeman will have _____ CDs, and Iesha will have _____ CDs.
Plan	Pick a strategy. One strategy is to make a list.
Solve	List the factors of each number.

_____ : _____

_____ : _____

The greatest common factor is _____.

$$\frac{\square}{\square} = \frac{\square \div \square}{\square \div \square} = \frac{\square}{\square}$$

The ratio of Freeman's CDs to Iesha's CDs is _____.

Check	Look over your solution. Did you answer the question?

Problem-Solving Strategies
☐ Draw a diagram.
☐ Use logical reasoning.
☐ Solve a simpler problem.
☑ Make a list.
☐ Work backward.

Copyright © Glencoe/McGraw-Hill, a division of The McGraw-Hill Companies, Inc.

10 **FOOTBALL** In the high school playoffs, the Creekville Cougars played the Southtown Sharks. The Cougars' season record was 12 wins and 4 losses. The Sharks' season record was 16 wins and 0 losses. What was the ratio of wins for the Cougars to wins for the Sharks? Check off each step.

_____ Understand: I underlined key words.

_____ Plan: To solve this problem, I will _____.

_____ Solve: The answer is _____.

_____ Check: I checked my answer by _____.

11 **CHESS** Jena and Niles played 10 games of chess. Jena won 6 of them. Write a ratio of Jena's wins to the total number of games in simplest form. _____

12 **Reflect** What is a greatest common factor? Explain using examples.

_____

_____

_____

▶ Skills, Concepts, and Problem Solving

Use the diagram to write each ratio as a fraction in simplest form.

13 circles and triangles to squares and pentagons _____

14 figures that are *not* triangles to total figures _____

15 triangles to circles and squares _____

Write the ratio of the length of the base to the height of the triangle as a fraction in simplest form.

16  height = 4 m _____

base = 16 m

17 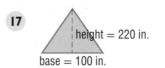 height = 220 in. _____

base = 100 in.

Copyright © Glencoe/McGraw-Hill, a division of The McGraw-Hill Companies, Inc.

Write the ratio of the length of the base to the height of each triangle as a fraction in simplest form.

18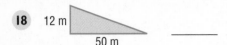
12 m
50 m _____

19
11 mm
7 mm _____

GRADES The quiz average is the ratio of the number of correct answers to the total number of exercises. Refer to the table to answer Exercises 20–22.

20 Which students had the same quiz average?

_____

What is that quiz average?

_____

Quiz Results		
Name	**Correct Answers**	**Number of Exercises**
Tabitha	27	36
Elvin	30	48
Adelfo	54	72
Emil	32	40
Geoff	33	44

21 Did the student with the most correct answers have the highest quiz average? Explain.

_____

_____

22 Explain the meaning of Adelfo's quiz average.

_____

Vocabulary Check **Write the vocabulary word that completes each sentence.**

23 A(n) _____ compares two quantities.

24 The _____ of 30 and 42 is 6.

25 **Writing in Math** Write the ratio of *5 pens out of a total of 6 pens* four different ways.

_____

Solve.

26 **GAMES** Nadia and Brock played 8 games of checkers. Nadia won 6 of them. Write a ratio of Nadia's wins to the total number of games in simplest form. _____

STOP

Copyright © Glencoe/McGraw-Hill, a division of The McGraw-Hill Companies, Inc.

Rates and Unit Costs

KEY Concept

The table shows the distance Marcus drove each day. A **rate** is a **ratio** of two measurements having different units.

Day	Monday	Tuesday	Wednesday	Thursday	Friday
Distance	60 mi	60 mi	60 mi	60 mi	60 mi

The rate $\dfrac{300 \text{ miles}}{5 \text{ days}}$ describes both the units of distance and time.

To show the **unit rate**, simplify the rate so that it has a denominator of 1 unit.

$$\frac{300 \text{ miles} \div 5}{5 \text{ days} \div 5} = \frac{60 \text{ miles}}{1 \text{ day}} \qquad \text{The unit rate is } \frac{60 \text{ miles}}{1 \text{ day}}.$$

Unit cost is the cost of a single item or unit of measure. The cost of a 20-ounce bottle of water is \$1.60.

$$\frac{\$1.60}{20 \text{ ounces}} = \frac{160 \text{ cents}}{20 \text{ ounces}} \rightarrow 20\overline{)160}^{\,8} \rightarrow \frac{8 \text{ cents}}{1 \text{ ounce}}$$

One ounce costs \$0.08.

VOCABULARY

rate
   a ratio comparing two quantities with different kinds of units

ratio
   a comparison of two numbers by division

unit cost
   the cost of a single item or unit

unit rate
   a rate that has a denominator of 1

Rates are often written using abbreviations, such as 300 mi/5 days, 60 mi/h, or \$0.21/oz.

Example 1

Andrés bought a pack of trading cards for \$6.65. The pack contains 15 cards. Find the unit cost to the nearest cent.

1. Write the rate as a fraction.
$$\frac{\$6.65}{15 \text{ cards}}$$

2. Divide the numerator by the denominator.

3. Round to the nearest cent.

$$\begin{array}{r} 0.44 \\ 15\overline{)6.65} \\ -60 \\ \hline 65 \\ -60 \\ \hline 5 \end{array}$$

Each card costs about \$0.44.

YOUR TURN!

Elise bought a bouquet of flowers for \$26.95. The bouquet contains 12 flowers. Find the unit cost to the nearest cent.

1. Write the rate as a fraction.

_____

2. Divide the numerator by the denominator.

3. Round to the nearest cent.

Each flower costs about _____.

GO ON

Copyright © Glencoe/McGraw-Hill, a division of The McGraw-Hill Companies, Inc.

Example 2

Find the unit rate for swimming 300 meters in 6 minutes. Use the unit rate to find the number of meters swam in 5 minutes.

1. Write the rate as a fraction.

$$\frac{300 \text{ meters}}{6 \text{ minutes}}$$

2. Find an equivalent rate with a denominator of 1.

   Divide the numerator and denominator by 6.

$$\frac{300 \div 6}{6 \div 6} = \frac{50}{1}$$

3. The unit rate is 50 meters/minute.

4. To find how many meters will be swam at this rate in 5 minutes, multiply the numerator and denominator by 5.

$$\frac{50 \text{ meters} \cdot 5}{1 \text{ minute} \cdot 5} = \frac{250 \text{ meters}}{5 \text{ minutes}}$$

At this rate, 250 meters will be swam in 5 minutes.

YOUR TURN!

Find the unit rate for traveling 616 feet in 7 seconds. Use the unit rate to find the number of feet traveled in 12 seconds.

1. Write the rate as a fraction.

2. Divide the numerator and denominator by _____.

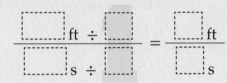

3. The unit rate is _____/_____.

4. Multiply the numerator and denominator by _____.

At this rate, _____ feet will be traveled in 12 seconds.

Copyright © Glencoe/McGraw-Hill, a division of The McGraw-Hill Companies, Inc.

Who is Correct?

Keli spent $2.00 to purchase 8 oranges. Find the unit rate.

Ashima

$$\frac{8 \div 2}{2 \div 2} = 4$$

Unit rate = $0.40/orange

Alana

$$\frac{2 \div 2}{8 \div 2} = \frac{1}{4}$$

$\frac{1}{4}$ of $1 is $0.25.

Unit rate = $0.25/orange

Cole

$$\begin{array}{r} 0.25 \\ 8\overline{)2.00} \\ -16 \\ \hline 40 \\ -40 \\ \hline 0 \end{array}$$

Unit rate = $0.25/orange

Circle correct answer(s). Cross out incorrect answer(s).

Guided Practice

Write each rate as a fraction. Find each unit rate.

1 140 words in 4 minutes

_____

2 12 books in 5 days

_____

Find each unit rate. Use the unit rate to find the unknown amount.

3 150 feet in 8 seconds; ☐ feet in 14 seconds

_____

4 $5 for 4 books; ☐ dollars for 15 books

_____

Step by Step Practice

5 Use the table to find which box of pens has the lowest unit cost.

Step 1 Find the unit cost of a 12-count package.

$\frac{5.76}{12}$ → 12)5.76 → $_____/pen

Box Size	Price
12-count	$5.76
16-count	$6.72
32-count	$11.84

Step 2 Find the unit cost of a 16-count package.

$\frac{\boxed{}}{\boxed{}}$ →)‾‾‾ → $_____/pen

Step 3 Find the unit cost of a 32-count package.

$\frac{\boxed{}}{\boxed{}}$ →)‾‾‾ → $_____/pen

Step 4 Which package costs the least per pen? _____

Which product has the lowest unit cost?

6 a 12-oz water bottle for $0.72 or a 24-oz water bottle for $1.92

12-oz bottle: $\frac{\boxed{}}{\boxed{}}$ →)‾‾‾ → $_____/oz

24-oz bottle: $\frac{\boxed{}}{\boxed{}}$ →)‾‾‾ → $_____/oz

The _____ water bottle costs less per ounce.

GO ON →

Copyright © Glencoe/McGraw-Hill, a division of The McGraw-Hill Companies, Inc.

Which product has the lowest unit cost?

7 a 6-pack of fruit bar for $1.98 or a 12-pack of fruit bar for $3.48

_____

8 a 16-oz box of rice for $3.04, a 32-oz box of rice for $3.84, or a 48-oz box of rice for $5.28

_____

9 50-count vitamins for $5.50, 100-count vitamins for $9.00, or 150-count vitamins for $15.00

_____

Step by Step Problem-Solving Practice

Copyright © Glencoe/McGraw-Hill, a division of The McGraw-Hill Companies, Inc.

Problem-Solving Strategies
☐ Draw a diagram.
☐ Look for a pattern.
☐ Guess and check.
☐ Make a table.
☑ Solve a simpler problem.

Solve.

10 **NATURE** A gray wolf can travel at speeds up to 27 kilometers in 25 minutes. A certain gazelle can travel at speeds up to 20 kilometers in 15 minutes. Which animal can travel at a faster rate?

Understand Read the problem. Write what you know.

A gray wolf can travel

_____ kilometers in _____ minutes.

A gazelle can travel

_____ kilometers in _____ minutes.

Plan Pick a strategy. One strategy is to solve a simpler problem. Find each unit rate.

Solve Write each unit rate as a fraction. Find an equivalent rate with a denominator of 1.

Wolf: $\dfrac{27 \div \rule{1cm}{0.4pt}}{25 \div \rule{1cm}{0.4pt}} = \rule{1.5cm}{0.4pt}$

Gazelle: $\dfrac{20 \div \rule{1cm}{0.4pt}}{15 \div \rule{1cm}{0.4pt}} = \rule{1.5cm}{0.4pt}$

Compare the unit rates.

_____ km/min < _____ km/min

The _____ travels at a faster rate.

Check Look over your answer. Did you answer the question?

11 **BUSINESS**   Mitch worked at a landscaping company for the summer. He earned $753 in 12 weeks. Justine worked for a grocery store for the summer. She earned $632.50 in 10 weeks. Find the unit rates to describe their weekly wages. Whose weekly wages were higher? Check off each step.

_____ Understand: I underlined key words.

_____ Plan: To solve the problem I will _____.

_____ Solve: The answer is _____

_____.

_____ Check: I checked my answer by _____.

12 **POPULATION**   The population of Pennsylvania is about 12.4 million people. Its land area is approximately 44,817 square miles. Find the population per square mile. _____

13 **Reflect**   Explain the difference between a rate and ratio. What is the difference between unit rate and unit cost?

_____

_____

Skills, Concepts, and Problem Solving

Write each rate as a fraction. Find each unit rate.

14 33 hits out of 40 at-bats

15 12 bars of soap for $9

_____   _____

Find each unit rate. Use the unit rate to find the unknown amount.

16 75 meters in 4 seconds; ☐ meters in 14 seconds _____

17 50 fence posts every 8 yards; ☐ fence posts in 20 yards _____

18 $15 for 8 pounds; ☐ dollars for 6 pounds _____

Which product has the lower unit cost? Round to the nearest cent.

19 4 cards for $3 or 10 cards for $8.50 _____

20 16 oz of glass cleaner for $3 or 8 oz of glass cleaner for $1.75 _____

GO ON

Copyright © Glencoe/McGraw-Hill, a division of The McGraw-Hill Companies, Inc.

Solve.

21 **SALES** Amanda sold 225 DVDs in 6 days, while Hayden sold 181 DVDs in 4 days. Who sold DVDs at a faster rate? Explain.

_____

22 **POPULATION** Which state has the lower population per square mile?

State	Population	Area in Square Miles
Montana	944,632	145,552
North Dakota	635,867	68,976

_____

_____

Vocabulary Check **Write the vocabulary word that completes each sentence.**

23 A ratio of two measurements or amounts of different units, where the denominator is 1 is a(n) _____.

24 The cost of a single item or unit is the _____.

25 **Writing in Math** Explain how to find the unit cost for a box of 8 granola bars for $3.36. Then determine the cost of buying 11 granola bars with the same unit cost.

_____

_____

 Spiral Review

Use the diagram shown at the right to write each ratio as a fraction in simplest form. (Lesson 5-1, p. 180)

26 The number of red tiles to the number of yellow tiles is _____.

27 The number of red tiles to the total number of tiles is _____.

28 The number of yellow tiles to the total number of tiles is _____.

STOP

Copyright © Glencoe/McGraw-Hill, a division of The McGraw-Hill Companies, Inc.

Use the diagrams to write each ratio as a fraction in simplest form.

1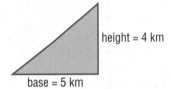

blue to yellow squares _____

2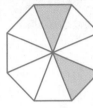

green parts to total parts _____

Write the ratio of the height to the length of the base of the triangle as a fraction in simplest form.

3 _____

height = 4 km

base = 5 km

4 _____

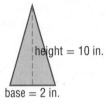

height = 10 in.

base = 2 in.

Write each rate as a fraction. Find each unit rate.

5 90 meters in 18 minutes _____

6 6 people in 150 miles _____

Write each ratio as a fraction in simplest form.

7 21 out of 147 boys wore glasses _____

8 15 black dogs out of 36 dogs _____

Which product has the lowest unit cost?

9 12-oz can for $1.56, a 16-oz can for $2.40, or a 32-oz can for $3.84

_____

10 9 lemons for $1.35, 14 lemons for $2.25, or 20 lemons for $3.80

_____

Solve.

11 **SPELLING**   Write a fraction in simplest form for the ratio of the number of vowels in *Math Triumphs* to the total number of letters. _____

12 **SPORTS**   Nick jumped rope 72 times in 48 seconds. What is his unit rate?

_____

Copyright © Glencoe/McGraw-Hill, a division of The McGraw-Hill Companies, Inc.

Proportions

KEY Concept

An equation stating that two **ratios** are equivalent is a **proportion**. For two ratios to form a proportion, their cross products must be equal.

$1 \cdot 6$ is one cross product.

$2 \cdot 3$ is the other cross product.

$$1 \cdot 6 = 2 \cdot 3$$
$$6 = 6 \qquad \text{The } \textbf{cross products} \text{ are equal.}$$

Cross multiplying works because a common denominator of the two fractions is $6 \cdot 2$ or 12. Multiply both sides of the proportion by $6 \cdot 2$.

On the left side cancel the 2s.
On the right side cancel the 6s.

$$(\cancel{2} \cdot 6) \frac{1}{\cancel{2}} = \frac{3}{\cancel{6}} (2 \cdot \cancel{6})$$

This results in $1 \cdot 6 = 2 \cdot 3$. Notice that $1 \cdot 6$ and $2 \cdot 3$ are the same as the two cross products of the original proportion.

Cross multiply to solve proportions when one value in the proportion is not known. If two bagels cost $0.50, you can use a proportion to find the cost of 12 bagels.

$$\frac{0.50}{2} = \frac{y}{12} \qquad y = \text{the cost of 12 bagels}$$
$$0.50 \cdot 12 = 2 \cdot y \qquad \text{Cross multiply.}$$
$$\frac{6}{2} = \frac{2y}{2} \qquad \text{Divide by 2 to find the value of } y.$$
$$3 = y \qquad \text{12 bagels would cost \$3.}$$

VOCABULARY

cross product
in a proportion, a cross product is the product of the numerator of one ratio and the denominator of the other ratio

proportion
an equation stating that two ratios are equivalent

ratio
a comparison of two quantities by division

Example 1

Determine whether the ratios are proportional. Write = or ≠ in the circle.

$\frac{3}{4} \bigcirc \frac{7}{21}$

1. Find the cross products.

2. The cross products are not equal. Therefore, the ratios are not proportional.

$3 \cdot 21 \bigcirc 4 \cdot 7$
$63 \neq 28$

Copyright © Glencoe/McGraw-Hill, a division of The McGraw-Hill Companies, Inc.

Copyright © Glencoe/McGraw-Hill, a division of The McGraw-Hill Companies, Inc.

YOUR TURN!

Determine whether the ratios are proportional. Write = or ≠ in the circle.

$\frac{2}{5} \bigcirc \frac{4}{10}$

1. Find the cross products.

2. The cross products are _____. Therefore, the ratios _____ proportional.

_____ $\bigcirc$ _____

_____

Example 2

Solve for *t*.

$\frac{14}{t} = \frac{10}{11}$

1. Find the cross products.

2. Solve.

$14 \cdot 11 = t \cdot 10$

$154 = 10t$

$t = 15.4$

YOUR TURN!

Solve for *w*.

$\frac{w}{6} = \frac{2.8}{7}$

1. Find the cross products.

2. Solve.

$w \cdot 7 =$ _____ $\cdot$ _____

$7w =$ _____

$w =$ _____

Example 3

Wyley Auto Sales orders 8 cars to every 3 trucks ordered. How many cars are ordered when 15 trucks are ordered?

1. Write the ratio of cars to trucks. $\frac{8}{3}$

> Do you remember how to write a ratio?
> Example: There are 3 girls in every 5 students.
> The ratio of girls to students is $\frac{3}{5}$.

2. Write another ratio of cars to trucks using 15 for the number of trucks. $\frac{c}{15}$

3. Write these two ratios as a proportion.

4. Cross multiply and solve.

cars → $\frac{8}{3} = \frac{c}{15}$ ← cars
trucks →   ← trucks

$\frac{8}{3} = \frac{c}{15}$

$8 \cdot 15 = 3c$

$\frac{120}{3} = \frac{3c}{3}$

$40 = c$

Forty cars are ordered when 15 trucks are ordered.

GO ON

YOUR TURN!

A recipe that makes 3 dozen cookies calls for 7 cups of flour. How many dozens of cookies can be made with 28 cups of flour?

1. Write the ratio of dozens of cookies to cups of flour. _____

2. Write another ratio of _____ to _____

   using _____ for the number of cups of flour. _____,

   where _____ is the number of dozens made with 28 cups of flour.

3. Write these two ratios as a proportion. _____ = _____

4. Cross multiply and solve.

$$\frac{\rule{2cm}{0.4pt}}{} = \frac{\rule{2cm}{0.4pt}}{}$$

$$\frac{\rule{2cm}{0.4pt}}{} \cdot \frac{\rule{2cm}{0.4pt}}{} = \frac{\rule{2cm}{0.4pt}}{}$$

$$\frac{\rule{2cm}{0.4pt}}{} = \frac{\rule{2cm}{0.4pt}}{}$$

$$f = \rule{2cm}{0.4pt}$$

_____ dozen cookies can be made with 28 cups of flour.

Who is Correct?

Irina uses 4 inches of wire for every 3.6 feet of ribbon to make big bows. If she has 48 inches of wire, how many feet of ribbon does she need to use all the wire?

Kraig

$\dfrac{4}{3.6} = \dfrac{48}{z}$

$4z = 48 \cdot 3.6$

$4z = 1{,}728$

$z = 432$

Irina needs 432 feet of ribbon.

Rasha

$\dfrac{4}{3.6} = \dfrac{z}{48}$

$192 = 3.6z$

$z = 53.3$

Irina needs 53.3 feet of ribbon.

Frances

$\dfrac{4}{3.6} = \dfrac{48}{z}$

$4z = 48 \cdot 3.6$

$4z = 172.8$

$z = 43.2$

Irina needs 43.2 feet of ribbon.

Circle correct answer(s). Cross out incorrect answer(s).

Copyright © Glencoe/McGraw-Hill, a division of The McGraw-Hill Companies, Inc.

▶ Guided Practice

Determine whether each pair of ratios is proportional. Write = or ≠ in each circle.

1 $\frac{5}{9}$ ◯ $\frac{10}{18}$

2 $\frac{2}{7}$ ◯ $\frac{18}{42}$

3 $\frac{36}{12}$ ◯ $\frac{12}{4}$

4 $\frac{350}{1,750}$ ◯ $\frac{2}{10}$

Step by Step Practice

5 Solve the proportion. $\frac{2}{12} = \frac{a}{36}$

Step 1 Cross multiply. _____

Step 2 Solve. _____

Step 3 The solution is _____. _____

Solve each proportion.

6 $\frac{n}{8} = \frac{3}{4}$

$= \dfrac{\underline{\hspace{1cm}} \cdot \underline{\hspace{1cm}}}{\underline{\hspace{1cm}} \cdot \underline{\hspace{1cm}}}$

$n = $ _____

7 $\frac{w}{3} = \frac{34}{51}$

$= \dfrac{\underline{\hspace{1cm}} \cdot \underline{\hspace{1cm}}}{\underline{\hspace{1cm}} \cdot \underline{\hspace{1cm}}}$

$w = $ _____

8 $\frac{9}{15} = \frac{b}{10}$ _____

9 $\frac{3}{5} = \frac{0.2}{d}$ _____

10 $\frac{2.5}{14.52} = \frac{7.5}{x}$ _____

11 $\frac{3}{n} = \frac{27}{18}$ _____

12 Two apples cost \$0.58. How many apples could you buy for \$2.32? _____

13 Refer to the price sticker on the books to the right.
How many books cost \$17.85? _____

14 Twenty bows make 8 centerpieces. How
many bows make 30 centerpieces? _____

15 Thirty-two yards of cloth make 6 blankets.
How many yards make 9 blankets? _____

GO ON ➤

Copyright © Glencoe/McGraw-Hill, a division of The McGraw-Hill Companies, Inc.

Step by Step Problem-Solving Practice

Solve.

Problem-Solving Strategies

☑ Make a table.

☐ Look for a pattern.

☐ Guess and check.

☐ Solve a simpler problem.

☐ Work backward.

16 PETS   Cats drink about 2 milliliters of water for each gram of food they eat. If a cat eats about 9 grams of food, how much water will it drink?

Understand   Read the problem. Write what you know.

Cats drink _____ milliliters of water for every _____ gram of food.

Plan   Pick a strategy. One strategy is to make a table.

water (mL)	2	4	6					
food (g)	1	2	3					

Solve   What pattern do you observe? Write the pattern as a ratio.

_____

Use the table or the ratio to find the solution.

_____

Check   Use your answer to write a ratio. Is your ratio equivalent to the ratio in the table?

17 BIKING   A group of bicyclers went on a bike ride. After 1 hour, they had traveled 16 miles. How long will it take for them to complete 60 miles?

Check off each step.

_____ Understand: I underlined key words.

_____ Plan: To solve this problem I will _____.

_____ Solve: The answer is _____.

_____ Check: I checked my answer by _____.

18 SCHOOL   On a field trip there must be 3 adults for every 24 students. If there are 12 adults, how many students can go on the field trip? _____

Copyright © Glencoe/McGraw-Hill, a division of The McGraw-Hill Companies, Inc.

19 Reflect   Describe a situation in which you might use a proportion.

_____

_____

▶ Skills, Concepts, and Problem Solving

Determine whether each pair of ratios is proportional. Write = or ≠ in each circle.

20 $\frac{2}{9} \bigcirc \frac{4}{16}$

21 $\frac{5}{10} \bigcirc \frac{4}{8}$

22 $\frac{3}{4} \bigcirc \frac{12}{3}$

23 $\frac{72}{12} \bigcirc \frac{12}{2}$

Solve each proportion.

24 $\frac{2}{5} = \frac{8}{x}$ $x =$ _____

25 $\frac{2}{7} = \frac{4}{y}$ $y =$ _____

26 $\frac{3}{5} = \frac{b}{30}$ $b =$ _____

27 $\frac{2}{9} = \frac{c}{36}$ $c =$ _____

28 $\frac{4}{5} = \frac{d}{25}$ $d =$ _____

29 $\frac{20}{4} = \frac{10}{f}$ $f =$ _____

30 $\frac{d}{16} = \frac{3}{8}$ $d =$ _____

31 $\frac{1.2}{9} = \frac{c}{1.5}$ $c =$ _____

Solve.

32 READING   Pravat read 4 pages in 6 minutes. At this rate, how long would it take him to read 6 pages?

_____

33 FOOD   How many Calories would you expect to find in 5 slices of the same kind of pizza? (See the photo at right.)

_____

34 BUSINESS   Jermaine can type 180 words in 3 minutes. How many words would you expect him to type in 10 minutes?

_____

2 slices = 575 Calories

GO ON

Copyright © Glencoe/McGraw-Hill, a division of The McGraw-Hill Companies, Inc.

Vocabulary Check **Write the vocabulary word that completes each sentence.**

35 A(n) _____ is an equation stating that two ratios or rates are equivalent.

36 A(n) _____ is the product of a numerator of one ratio and the denominator of another ratio.

37 **Writing in Math** Describe the relationship between the two ratios in a proportion.

_____

_____

▶ Spiral Review

Use the diagram to write each ratio as a fraction in simplest form. (Lesson 5-1, p. 180)

38 apples and plums to pears and bananas _____

39 pieces of fruit that are not bananas to total fruit _____

40 apples to plums and bananas _____

Write each rate as a fraction. Find each unit rate. (Lesson 5-2, p. 187)

41 5 pounds of turkey for 8 people? _____

42 168 miles in 3 hours _____

Solve. (Lesson 5-2, p. 187)

43 **GAS MILEAGE** Jude can drive 330 miles on 12 gallons of gas. Find the unit rate. Use the unit rate to find the number of miles Jude can drive on 64 gallons of gas.

_____

_____

44 **POPULATION** The population of New Mexico is about 1,954,599. Its land area is about 121,356 square miles. Find the estimated population per square mile.

_____

STOP

Copyright © Glencoe/McGraw-Hill, a division of The McGraw-Hill Companies, Inc.

Solve Problems Using Proportions

KEY Concept

Similar figures have the same shape but may have different sizes. The corresponding angles of similar figures are congruent, or the same. The corresponding sides of similar figures are proportional.

The corresponding angles are
∠ *B* and ∠ *E*
∠ *A* and ∠ *D*
∠ *C* and ∠ *F*

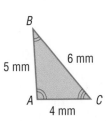

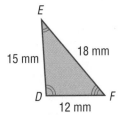

The corresponding sides are
AB and *DE*
BC and *EF*
CA and *FD*

The corresponding angles of triangles *ABC* and *DEF* are congruent. The ratios of the corresponding sides are equivalent.

$$\frac{18}{6} = \frac{12}{4} = \frac{15}{5} = \frac{3}{1} = 3$$

If two figures are similar, the ratio of two sides of one figure is equal to the ratio of the corresponding two sides of the other figure. In triangles *ABC* and *DEF*, the ratio $\frac{AB}{AC}$, which is $\frac{5}{4}$, is equal to the ratio $\frac{DE}{DF}$, which is $\frac{15}{12}$.

VOCABULARY

proportion
an equation stating that two ratios or rates are equivalent

similar figures
figures that have the same shape but may have different sizes

Proportions can be used to find unknown measures in similar figures, in percent proportions, and for unit conversions.

Example 1

Find the value of x. Triangle GHI is similar to triangle JKL.

1. The ratio of corresponding sides $\overline{GH}$ and $\overline{JK}$ is $\frac{9}{21}$.

2. The ratio of corresponding sides $\overline{HI}$ and $\overline{KL}$ is $\frac{x}{14}$.

3. Since the two triangles are similar, these two ratios are equal. Write the proportion and solve for *x*.

4. The length of side $\overline{HI}$ is 6 meters.

$$\frac{9}{21} = \frac{x}{14}$$

$21x = 14(9)$ Find the cross products.

$$\frac{21x}{21} = \frac{126}{21}$$ Simplify. Divide by 21.

$x = 6$

Be careful to write the ratios in the same order.

Copyright © Glencoe/McGraw-Hill, a division of The McGraw-Hill Companies, Inc.

Find the value of *x*. Triangle *MNO* is similar to triangle *PQR*.

1. The ratio of corresponding sides $\overline{MN}$ and $\overline{PQ}$ is _____.

2. The ratio of corresponding sides $\overline{NO}$ and $\overline{QR}$ is _____.

3. Since the two triangles are similar, these two ratios are equal.

    Write the proportion and solve for *x*.

4. The length of side $\overline{PQ}$ is _____.

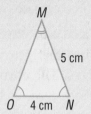

Example 2

An A-37 Dragonfly jet can travel 42 miles in 4 seconds. How long would it take the jet to travel a total of 105 miles?

1. Write a ratio for miles to seconds.

$$\frac{42 \text{ mi}}{4 \text{ s}}$$

2. Set up a proportion to find the time it would take to travel 105 miles.

$$\frac{42 \text{ mi}}{4 \text{ s}} = \frac{105 \text{ mi}}{t}$$

3. Cross multiply and solve.

$$42t = 4 \cdot 105$$
$$42t = 420$$

$$t = 10$$

4. It will take **10 seconds** for the jet to travel 105 miles.

Copyright © Glencoe/McGraw-Hill, a division of The McGraw-Hill Companies, Inc.

Copyright © Glencoe/McGraw-Hill, a division of The McGraw-Hill Companies, Inc.

YOUR TURN!

A cargo airplane is traveling 987 miles. It has flown 282 miles in 1.5 hours. How long will it take the airplane to fly the entire trip if the plane travels at the same rate of speed?

1. Write a ratio for miles to hours.   _____

2. Set up a proportion to find the time it would take to travel 987 miles.   _____ = _____

3. Cross multiply and solve.   _____

_____

_____

4. It will take _____ hours for the airplane to complete its journey.

Who is Correct?

Edita bicycled for 25 minutes and burned 293 calories. Use a proportion to find how long it would take her to burn 704 calories at that rate.

Marie

$\dfrac{293 \text{ Cal}}{25 \text{ min}} = \dfrac{704 \text{ Cal}}{x}$

$293x = 25 \cdot 704$

$x = 60.07 \text{ min}$

Liam

$\dfrac{704 \text{ Cal}}{25 \text{ min}} = \dfrac{293 \text{ Cal}}{x}$

$704x = 25 \cdot 293$

$x = 10.40 \text{ min}$

Alejandro

$\dfrac{293 \text{ Cal}}{25 \text{ min}} = \dfrac{x}{704 \text{ Cal}}$

$293 \cdot 704 = 25x$

$x = 8,250$

Circle correct answer(s). Cross out incorrect answer(s).

▶ Guided Practice

Find the value of x in each pair of similar triangles.

1 $x =$ _____

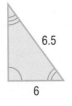

2 $x =$ _____

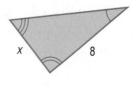
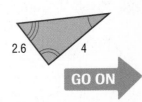

3 An Apache helicopter traveled 14.2 kilometers in 3 minutes. If the pilot maintains this rate of speed, how far will the Apache travel in 20 minutes?

Step 1 Write a ratio for kilometers to minutes.

$$\dfrac{km}{min}$$

Step 2 Set up a proportion to find the kilometers the Apache would travel in 20 minutes.

$$\dfrac{km}{min} = \dfrac{km}{min}$$

Step 3 Cross multiply and solve.

$$=$$

Step 4 The Apache could travel about _____ kilometers in 20 minutes.

Solve.

4 **WORK** Carson earns $35 for working 5 hours. At that rate, how many hours would he need to work in order to earn $140?

$$\dfrac{}{} = \dfrac{}{}$$

$$\dfrac{}{} = \dfrac{}{}$$

$$x =$$

Carson needs to work _____ hours.

5 Miranda wants to attend a class trip to the ballet. The trip will cost $65. She has already saved $25. She can earn $4 an hour cleaning the basement for her grandmother. How many hours will she need to clean in order to pay for the trip?

_____

6 Presta's mom drove for 2 hours at 30 miles per hour through city traffic. Then she drove 3 more hours at 55 miles per hour on the highway. How far did she drive?

_____.

Copyright © Glencoe/McGraw-Hill, a division of The McGraw-Hill Companies, Inc.

Step by Step Problem-Solving Practice

Solve.

Problem-Solving Strategies
- ☑ Use a table.
- ☐ Look for a pattern.
- ☐ Guess and check.
- ☐ Solve a simpler problem.
- ☐ Act it out.

7 **GEOMETRY** Each side of triangle ABC is $3\frac{1}{4}$ times as long as the corresponding side of triangle FGH. Find the perimeter of triangle ABC.

Understand Read the problem. Write what you know.

Each side of triangle ABC is _____ times as long as the corresponding side of triangle FGH.

Plan Pick a strategy. One strategy is to make a table showing the corresponding sides.

Solve First, fill in the corresponding sides of triangles FGH and ABC. Then, fill in the measurements of the sides of triangle FGH. All measurements are in inches.

Multiply the length of each side of triangle FGH by $3\frac{1}{4}$ or 3.25 to find the lengths of the sides of triangle ABC. Complete the table.

	side	length	side	length	side	length
△FGH	$\overline{FG}$	3	$\overline{GH}$	5	$\overline{HF}$	2
△ABC	$\overline{AB}$					

The perimeter is the sum of the lengths of the sides. What is the perimeter of triangle ABC? _____

Check Does the answer make sense? Look over your solution. Did you answer the question?

8 **SHADOWS** At the same time of day, the height of different objects and their shadows are proportional. If a 6-ft-high storage shed casts a shadow 5 feet long, how tall is a tree that casts a 12.5-ft shadow? Check off each step.

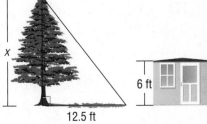
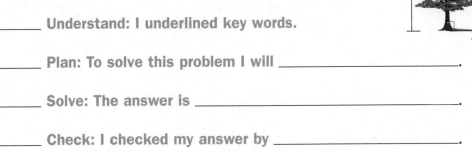

_____ **Understand: I underlined key words.**

_____ **Plan: To solve this problem I will** _____.

_____ **Solve: The answer is** _____.

_____ **Check: I checked my answer by** _____.

GO ON

Copyright © Glencoe/McGraw-Hill, a division of The McGraw-Hill Companies, Inc.

9 **EVENTS** Eloise sells pizza slices at a carnival. She sold 180 slices and then took a 5-minute break. She then sold another 60 slices to complete the event. It took her 6 hours to complete the sale from start to finish. About how many pizza slices did she sell each hour?

_____

10 **Reflect** List the types of problems that proportions can be used to solve.

_____

_____

▶ Skills, Concepts, and Problem Solving

Find the value of *x* in each pair of similar figures.

11 $x =$ _____

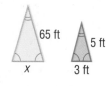

12 $x =$ _____

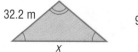

13 $x =$ _____

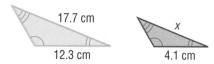

14 $x =$ _____

Use a proportion to solve.

15 Dante's remote control car can travel 3 miles in 18 minutes. At this speed, how long will it take to travel 25 miles?

_____

16 Halona canoed 3 miles in 29 minutes. At this rate, how long would it take her to canoe 5 miles?

_____

17 At 82°F, sound travels approximately 2,299 meters in 6.6 seconds through dry air. Find the speed of sound per second under these conditions.

_____

18 A Spine-tail swift can fly 31.5 miles in 18 minutes. How far would this bird fly if it continued at the same rate for 45 minutes?

_____

Copyright © Glencoe/McGraw-Hill, a division of The McGraw-Hill Companies, Inc.

Solve.

19 **SAFETY** Safe Child, Inc. produces 39 car seats every 2 days. How long will it take the company to produce 429 car seats?

_____

20 **FITNESS** Winston bicycled 5 miles on Saturday morning at a rate of 12.5 miles per hour. How many minutes did he bicycle?

_____

He bicycled 3 miles on Monday at a rate of 11.7 miles per hour. Did he bicycle for a longer amount of time on Saturday or Monday?

_____

Vocabulary Check **Write the vocabulary word that completes each sentence.**

21 Figures whose shapes are the same but may have different sizes are

_____.

22 A(n) _____ is a comparison of two numbers by division.

23 **Writing in Math** Ruth says that if two figures are similar, their corresponding sides are equal and their corresponding angles are proportional. Is she correct? Explain.

_____

_____

▶ **Spiral Review**

Solve. (Lesson 5-3, p. 194)

24 If a turtle travels 7.5 feet in 15 minutes, what is its rate per minute?

_____

25 If a motorcycle travels 30 miles per hour, what is the distance it travels in 5.5 hours? _____

26 **ENTERTAINMENT** Catalina sold 330 tickets to the theater in 5 hours while working at the ticket booth. Later, Troy sold 480 tickets while working an 8-hour shift. Who sold tickets at a higher rate? Explain. (Lesson 5-2, p. 187)

_____

_____

Copyright © Glencoe/McGraw-Hill, a division of The McGraw-Hill Companies, Inc.

STOP

Determine whether each pair of ratios is proportional. Write = or ≠ in each circle.

1. $\dfrac{5}{2}$ ◯ $\dfrac{10}{5}$

2. $\dfrac{15}{5}$ ◯ 3

Solve each proportion.

3. $\dfrac{34}{15} = \dfrac{n}{5}$ $n =$ _____

4. $\dfrac{2}{3} = \dfrac{72}{d}$ $d =$ _____

5. $\dfrac{0.75}{5} = \dfrac{n}{25}$ $n =$ _____

6. $\dfrac{27}{3} = \dfrac{90}{c}$ $c =$ _____

Find the value of x in each pair of similar figures.

7. $x =$ _____

x   12 in.

5 in.   10 in.

8. $x =$ _____

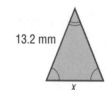

13.2 mm   3.3 mm

x   2.4 mm

Use a proportion to solve.

9. A child weighs about 27 kilograms. If 1 kilogram is about 2.2 pounds, how much does the child weigh in pounds?

_____

10. Two dozen books cost $26. How much do 3.5 dozen books cost?

_____

Solve.

11. **TRAVEL**   The gasoline tank of Nayla's car holds 14 gallons of gas. A gallon of gas costs $3.19. If Roxanna has $35, does she have enough money to fill the tank? Exactly how much will she spend if the tank is completely empty?

_____

12. **EARTH SCIENCE**   Surface waves from an earthquake travel about 3.7 miles per second through Earth's crust. How long would it take for a surface wave to travel 444 miles?

_____

Copyright © Glencoe/McGraw-Hill, a division of The McGraw-Hill Companies, Inc.

Vocabulary and Concept Check

proportion, *p. 194*
rate, *p. 187*
ratio, *p. 180*
similar figures, *p. 201*
unit cost, *p. 187*
unit rate, *p. 187*

Write the vocabulary word that completes each sentence.

1. An equation stating that two ratios are equivalent is a(n) _____.

2. A(n) _____ is a ratio of two measurements or amounts made with different units, such as 2 miles in 5 minutes.

3. The corresponding sides of _____ are proportional.

4. A(n) _____ is a comparison of two numbers by division.

Write the correct vocabulary term in each blank.

5. $4.19 per gallon _____

6. 67 miles per hour _____

Lesson Review

5-1 Ratios (pp. 180–186)

Write each ratio as a fraction in simplest form.

7. 16 people to 24 chairs _____

8. 10 pens to 30 pencils _____

9. 12 cell phones to 4 walkie talkies

_____

10. 84 buttons to 21 pairs of jeans

_____

Example 1

Write the ratio as a fraction in simplest form.
6 math books out of 30 total books

1. Write the ratio with the number of math books in the numerator and the total number of books in the denominator.

$$\frac{6}{30}$$

2. Write the fraction in simplest form.

$$\frac{6 \div 6}{30 \div 6} = \frac{1}{5}$$

Copyright © Glencoe/McGraw-Hill, a division of The McGraw-Hill Companies, Inc.

5-2 Rates and Unit Costs (pp. 187–192)

Write each rate as a fraction. Find each unit rate.

11. 228 jumps in 4 minutes

_____

12. 48 waves in 12 seconds

_____

13. 60 gallons in 5 minutes

_____

Example 2

Write the rate 60 rotations per 10 seconds as a fraction. Find the unit rate.

1. Write the rate as a fraction.

$$\frac{60 \text{ rotations}}{10 \text{ seconds}}$$

2. Find an equivalent rate with a denominator of 1.

$$\frac{60 \text{ rotations} \div 10}{10 \text{ seconds} \div 10} = \frac{6 \text{ rotations}}{1 \text{ second}}$$

3. Name the unit rate.

6 rotations per second or 6 rotations/s

5-3 Proportions (pp. 194–200)

Determine whether each pair of ratios is proportional. Write = or ≠ in each circle.

14. $\frac{3}{4}$ ◯ $\frac{21}{28}$

15. $\frac{1}{7}$ ◯ $\frac{6}{56}$

16. $\frac{5}{7}$ ◯ $\frac{60}{84}$

Solve each proportion.

17. $\frac{7}{9} = \frac{x}{45}$ _____

18. $\frac{p}{3} = \frac{12}{4.5}$ _____

19. $\frac{6}{7} = \frac{78}{m}$ _____

20. $\frac{2.2}{14} = \frac{t}{7}$ _____

Example 3

Determine whether the ratios are proportional. Write = or ≠ in the circle.

$$\frac{2}{3} \bigcirc \frac{26}{39}$$

1. Find the cross products.

$$2 \cdot 39 \bigcirc 3 \cdot 26$$
$$78 = 78$$

2. The cross products are equal. The ratios form a proportion.

Example 4

Solve for n.
$$\frac{7}{8} = \frac{n}{24}$$

1. Find the cross products. $7 \cdot 24 = 8n$

$$\frac{168}{8} = \frac{8n}{8}$$

2. Solve. $21 = n$

Copyright © Glencoe/McGraw-Hill, a division of The McGraw-Hill Companies, Inc.

5-4 Solve Problems Using Proportions (pp. 201–207)

Find the value of *x* in each pair of similar figures.

21 $x =$ _____

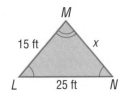

22 $x =$ _____

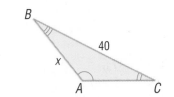

Use a proportion to solve.

23 A 12-oz package of beads contains 21 pieces. How many beads would you expect to find in a 16-oz package?

_____

24 Tyler jogged 6 miles in 36 minutes. How long will it take him to run 8 miles?

_____

25 The Concorde jet can travel 133.2 miles in 6 minutes. How far can the jet travel in 9 minutes? _____

Example 5

Find the value of *x* in the pair of similar figures.

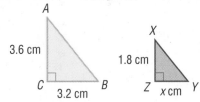

1. The ratio of sides *AC* to *CB* in triangle *ABC* is $\frac{3.6}{3.2}$.

2. The ratio of the corresponding side measures in triangle *XYZ* is $\frac{1.8}{x}$.

3. Set up a proportion and solve for *x*.

$$\frac{3.6}{3.2} = \frac{1.8}{x}$$
$$3.6x = 1.8(3.2)$$
$$\frac{3.6x}{3.6} = \frac{5.76}{3.6}$$

4. The length of $\overline{YZ}$ is 1.6 centimeters.

Example 6

Use a proportion to solve.

About 17 out of every 25 customers at The Craft Store purchase scrapbooking supplies. Sunday there were 75 customers. How many customers would purchase scrapbooking supplies at this rate?

1. Write a proportion. Use *n* to represent the number of scrapbooking customers. $\frac{17}{25} = \frac{n}{75}$

2. Solve for *n*. $17(75) = 25n$

$$1{,}275 = 25n$$

3. On Sunday, about 51 customers would purchase scrapbooking supplies. $\frac{1275}{25} = \frac{25n}{25}$

$$51 = n$$

Copyright © Glencoe/McGraw-Hill, a division of The McGraw-Hill Companies, Inc.

Use the diagram to write each ratio as a fraction in simplest form.

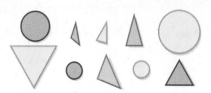

1 blue figures to total figures

2 triangles to circles

_____

_____

Write the ratio of the length of the base to the height of each triangle as a fraction in simplest form.

3

_____

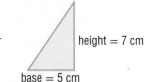

height = 7 cm

base = 5 cm

4

_____

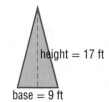

height = 17 ft

base = 9 ft

Write each rate as a fraction. Find each unit rate.

5 180 pages in 3 books

6 12 ounces in 3 bottles

_____

_____

Write each ratio as a fraction in simplest form.

7 18 out of 144 were not wearing jeans _____

8 7 of the 25 children in the class play soccer _____

9 112 markers in 16 bags _____

Which product has the lowest unit cost? Round to the nearest cent.

10 8-oz bag of crackers for $1.99, a 12-oz bag of crackers for $2.49, or a 16-oz bag of crackers for $2.99

_____

11 4 limes for $1, 10 limes for $2, or 24 limes for $6

_____

12 4 flower pots for $59.99, 6 flower pots for $74.99, or 10 flower pots for $99.99

_____

Copyright © Glencoe/McGraw-Hill, a division of The McGraw-Hill Companies, Inc.

Determine whether the ratios are proportional. Write = or ≠ in each circle.

13 $\dfrac{3}{5}$ ◯ $\dfrac{2}{3}$

14 $\dfrac{5}{9}$ ◯ $\dfrac{15}{27}$

Solve each proportion.

15 $\dfrac{p}{18} = \dfrac{3}{2}$ $p =$ _____

16 $\dfrac{9}{25} = \dfrac{36}{f}$ $f =$ _____

Solve.

17 **WRITING**   Keith was working on his lab journal for his science class. He completed 144 of the book's 200 pages. Write the pages Keith has completed to the total number of pages as a ratio in simplest form.

_____

18 **TRAVEL**   Margo drove her van 186 miles in 3 hours. What was her unit rate?

_____

19 **ART**   The two triangles shown are proportional. What is the length of the missing side?

20 in. 18 in.   x 15 in.

_____

Correct the mistakes.

20 At the Johnson Family Market, a sign in the window read: "8-oz bag of peanuts for $2.99. That's less than $5 per pound!" Is the sign correct?

_____

_____

21 Jeremy's mother needed $1\frac{1}{2}$ pounds of walnuts for a recipe. She looked at the store's ad and saw that walnuts were on sale for $6.99 per pound. She gave Jeremy a 10-dollar bill to buy the walnuts. What mistake did she make?

_____

STOP

_____

Copyright © Glencoe/McGraw-Hill, a division of The McGraw-Hill Companies, Inc.

Select the best answer and fill in the corresponding circle on the sheet at right.

1 Brianna finished reading a novel in 8 days. The book was 384 pages. About how many pages did she read per day?

    A 48 pages   **C** 96 pages

    B 64 pages   **D** 112 pages

2 A store sells an 8-pack of juice for $4. What is the cost of one bottle of juice?

    A $0.50   **C** $2.00

    B $1.00   **D** $4.00

3 In Allie's bookshelf, there are 32 science fiction books, 10 nonfiction titles, and 4 biographies. What is the ratio of science fiction to nonfiction books?

    A $\frac{8}{1}$; 8:1, or 8 to 1

    B $\frac{16}{5}$; 16:5, or 16 to 5

    C $\frac{5}{16}$; 5:16, or 5 to 16

    D $\frac{16}{23}$; 16:23, or 16 to 23

4 Diego is an avid biker. He rides about 140 miles every 4 days. At this rate, how many miles does he ride in 6 days?

    A 35 miles   **C** 210 miles

    B 175 miles   **D** 840 miles

5 Mrs. Delgado and her 26 students are going to the theater. Admission and lunch for everyone will cost $364.50. What is the price per person?

    A $13.50   **C** $14.50

    B $14.02   **D** $15.00

6 Isabel is running in a 26.2-mile marathon. If she completes the marathon in 4 hours, what rate did she average?

    A 26.2 miles   **C** 5.15 miles/hour

    B 4 hours   **D** 6.55 miles/hour

7 Cecil is driving to Chicago, Illinois. He makes the 275-mile trip in 5 hours. What is Cecil's average speed?

    A 1,375 miles   **C** 55 miles/hour

    B 5 hours   **D** 65 miles/hour

8 Write a ratio that compares the number of circles to the number of squares.

    A 1 to 1   **C** 4 to 1

    B 1 to 4   **D** 1 to 5

Copyright © Glencoe/McGraw-Hill, a division of The McGraw-Hill Companies, Inc.

9 One inch equals about 2.54 centimeters. About how many inches equal 12.7 centimeters?

    A 3 inches   **C** 8 inches

    B 5 inches   **D** 32.258 inches

---

10 A horse galloped at 13 miles per hour for 2 hours. Then the horse galloped at 15 miles per hour for 3 hours. How far did the horse travel in all?

    A 26 miles   **C** 71 miles

    B 45 miles   **D** 195 miles

---

11 Triangle *ABC* is similar to triangle *DEF*. Find the value of *x*.

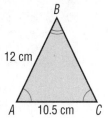

    A 6   **C** 7.75

    B 7.5   **D** 8

---

12 In Mr. Cameron's class, 5 out of 32 students play basketball. There are 128 students in the eighth grade. If the ratio is the same for the entire eighth grade, how many students play basketball?

    A 4   **C** 20

    B 5   **D** 32

Copyright © Glencoe/McGraw-Hill, a division of The McGraw-Hill Companies, Inc.

ANSWER SHEET

Directions: Fill in the circle of each correct answer.

1 Ⓐ Ⓑ Ⓒ Ⓓ
2 Ⓐ Ⓑ Ⓒ Ⓓ
3 Ⓐ Ⓑ Ⓒ Ⓓ
4 Ⓐ Ⓑ Ⓒ Ⓓ
5 Ⓐ Ⓑ Ⓒ Ⓓ
6 Ⓐ Ⓑ Ⓒ Ⓓ
7 Ⓐ Ⓑ Ⓒ Ⓓ
8 Ⓐ Ⓑ Ⓒ Ⓓ
9 Ⓐ Ⓑ Ⓒ Ⓓ
10 Ⓐ Ⓑ Ⓒ Ⓓ
11 Ⓐ Ⓑ Ⓒ Ⓓ
12 Ⓐ Ⓑ Ⓒ Ⓓ

Success Strategy

Easier questions usually come before harder ones. For the more difficult questions, try to break the information down into smaller pieces. Make sure the answer is reasonable and matches the question asked.

STOP

Squares, Square Roots, and the Pythagorean Theorem

You can use squared numbers and square roots to solve problems.

Malina's bedroom is in the shape of a square. She wants to cover the floor with tile. She will need to know how much area will be covered. If the length of the room equals 8 feet, what is the area of the room?

Copyright © Glencoe/McGraw-Hill, a division of The McGraw-Hill Companies, Inc.

STEP **1** Quiz

Math Online ▷ Are you ready for Chapter 6? Take the Online Readiness Quiz at *glencoe.com* to find out.

STEP **2** Preview

Get ready for Chapter 6. Review these skills and compare them with what you will learn in this chapter.

What You Know	What You Will Learn

What You Know

You know how to multiply the number of rows by the number of columns in an area model.

$$3 \cdot 4 = 12$$

TRY IT!

Write a multiplication equation for each area model.

1   **2**

_____   _____

You know how to locate points on a coordinate graph.

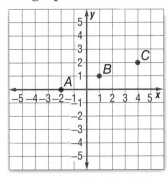

Point *C* is located at (4, 2). To locate Point *C*, follow the *x*-axis to the right 4 units, and follow the *y*-axis up 2 units.

TRY IT!

Name the location of each point on the coordinate grid.

3   Point *A* _____

4   Point *B* _____

What You Will Learn

Lesson 6-2

You can find **square roots** using square area models.

The model shows that $4 \cdot 4 = 16$. This can also be written as $4^2 = 16$.

To find a square root of 16, find the two equal factors of 16.

$$\sqrt{16} = \sqrt{4 \cdot 4} = 4$$

So, a square root of 16 is 4.

Lesson 6-5

You can connect the points on the grid to make a line. The steepness of the line is called its slope. You can measure the slope of the line.

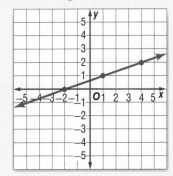

$$\text{Slope} = \frac{\text{rise}}{\text{run}} = \frac{\text{number of units up or down}}{\text{number of units left or right}} = \frac{1}{3}$$

The slope of the line that connects Points *A*, *B*, and *C* is $\frac{1}{3}$.

Copyright © Glencoe/McGraw-Hill, a division of The McGraw-Hill Companies, Inc.

Squaring a Number

KEY Concept

The multiplication sentences below have two identical **factors**.

$1 \cdot 1 = 1$	$5 \cdot 5 = 25$	$9 \cdot 9 = 81$
$2 \cdot 2 = 4$	$6 \cdot 6 = 36$	$10 \cdot 10 = 100$
$3 \cdot 3 = 9$	$7 \cdot 7 = 49$	$11 \cdot 11 = 121$
$4 \cdot 4 = 16$	$8 \cdot 8 = 64$	$12 \cdot 12 = 144$

Multiplication sentences can be modeled using a square array because the number of rows and the number of columns are the same. The **square of a number** is the product of two identical factors.

2^2 3^2 4^2 5^2
$2 \cdot 2$ $3 \cdot 3$ $4 \cdot 4$ $5 \cdot 5$
4   9   16   25

VOCABULARY

base
in a power, the number used as a factor; in 4^2, the base is 4

exponent
in a power, the number of time the base is used as a factor; in 4^2, the exponent is 2

factor
a number that is multiplied by another number

square of a number
the product of a number multiplied by itself; $4 \cdot 4$, or 4^2

The exponent 2 shows that the base is used as a factor 2 times. In the expression 5^2, 5 is the **base** and 2 is the **exponent**.

Example 1

Write an equation using exponents to represent the model.

1. How many rows in the array? 6

2. How many columns? 6

3. Label the array.

4. Write the expression and product. 6 • 6; 36

5. Write the equation using exponents. 6 is multiplied twice, so $6^2 = 36$.

6 • 6

Copyright © Glencoe/McGraw-Hill, a division of The McGraw-Hill Companies, Inc.

Copyright © Glencoe/McGraw-Hill, a division of The McGraw-Hill Companies, Inc.

YOUR TURN!

Write an equation using exponents to represent the model.

1. How many rows in the array? _____

2. How many columns? _____

3. Label the array.

4. Write the expression and product.

   _____ • _____ ; _____

5. Write the equation using exponents.

   _____ = _____

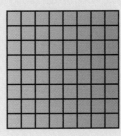

_____ • _____

Example 2

Evaluate the expression 4^2.

1. What number is the base? **4**

2. What number is the exponent? **2**

3. Multiply to find the value of the expression.
   $4 \cdot 4 = 16$

YOUR TURN!

Evaluate the expression 5^2.

1. What number is the base? _____

2. What number is the exponent? _____

3. Multiply to find the value of the expression. _____ • _____ = _____

Who is Correct?

Evaluate the expression 7^2.

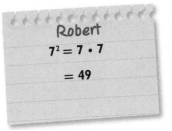

Camille
$7^2 = 7 \cdot 2$
$= 14$

Edward
$7^2 = 7 \cdot 7$
$= 14$

Robert
$7^2 = 7 \cdot 7$
$= 49$

Circle correct answer(s). Cross out incorrect answer(s).

▶ Guided Practice

Write an equation using exponents to represent each model.

1 _____ = _____

2 _____ = _____

Step by Step Practice

3 Evaluate the expression 9^2.

Step 1 What is the base? _____

Step 2 What is the exponent? _____

The exponent tells you how many times the _____ is used as a _____.

Step 3 Multiply to find the value of the expression. _____

Evaluate each expression.

4 10^2 base: _____

exponent: _____

$10^2 =$ _____ · _____ = _____

5 12^2 _____

6 8^2 _____

7 6^2 _____

8 3^2 _____

9 5^2 _____

10 11^2 _____

11 1^2 _____

12 4^2 _____

Copyright © Glencoe/McGraw-Hill, a division of The McGraw-Hill Companies, Inc.

Copyright © Glencoe/McGraw-Hill, a division of The McGraw-Hill Companies, Inc.

Step by Step Problem-Solving Practice

Solve.

Problem-Solving Strategies
☑ Draw a diagram.
☐ Use logical reasoning.
☐ Solve a simpler problem.
☐ Work backward.
☐ Make a table.

13 **INTERIOR DESIGN**   Rajesh is buying new carpet for his room. His room is square and has lengths of 12 feet. How many square feet of carpet will he need?

Understand   Read the problem. Write what you know.

The bedroom is a _____ with lengths of _____ feet.

Plan   Pick a strategy. One strategy is to draw a diagram.

You can make an array to represent his room.

Solve   Count the squares to find how many square feet fill the room.

The total number of squares is the same as the product of _____ • _____.

_____ • _____ = _____

So, Rajesh will need _____ square feet of carpet.

Check   Does your answer seem reasonable?
Count each square unit in the diagram.

14 **HEALTH**   Jocelyn has been training for a marathon. The first week she trained 1 hour, the second week she trained 4 hours, and the third week she trained 9 hours. If she continues this pattern, how many hours will she train for the marathon the fourth week? Check off each step.

_____ Understand: I underlined key words.

_____ Plan: To solve the problem, I will _____.

_____ Solve: The answer is _____.

_____ Check: I checked my answer by _____

_____.

GO ON

15 **SCIENCE** Derrick is conducting an experiment growing mold. He measures the area it covers daily. Complete the table showing the mold's growth.

Dimensions	1 • 1	2 • 2	3 •						
Exponential Form	1^2	2^2							
Area	1	4							

16 **Reflect** Define an exponent. Explain why you would use an exponent.

_____

_____

 Skills, Concepts, and Problem Solving

Write an equation using exponents to represent each model.

17 _____

18 _____

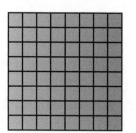

19 _____

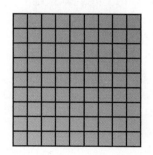

20 _____

Copyright © Glencoe/McGraw-Hill, a division of The McGraw-Hill Companies, Inc.

Write an equation using exponents to represent each model.

21 _____

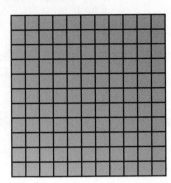

22 _____

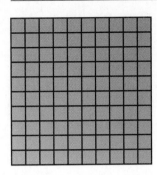

Evaluate each expression.

23 7^2 _____

24 12^2 _____

25 11^2 _____

26 9^2 _____

27 2^2 _____

28 8^2 _____

29 4^2 _____

30 6^2 _____

31 5^2 _____

32 10^2 _____

33 3^2 _____

34 1^2 _____

Solve.

35 **PUZZLES**   Write a two-digit number the sum of whose digits is 10.
The number is a squared number. _____

36 **PUZZLES**   Write a two-digit number the sum of whose digits is 7.
The number is a squared even number. _____

37 **PUZZLES**   Write a two-digit number the sum of whose digits is 9.
The number is a squared odd number. _____

GO ON

Copyright © Glencoe/McGraw-Hill, a division of The McGraw-Hill Companies, Inc.

38 FARMING Darin planted 12 rows of 12 bean plants. Write an expression using an exponent to represent the number of bean plants. Find the value of the expression.

_____

39 REUNION The reunion committee has arranged name tags on the welcome table in 13 rows of 13. Write an expression using an exponent to represent the number of name tags placed on the table. Find the value of the expression.

_____

40 OFFICE SPACE Sarasa is arranging cubicles for her company. She will use 10 rows of 10 cubicles. Make an array to represent the area.

Write an expression using an exponent to represent the number of cubicles Sarasa will arrange. Find the value of the expression.

_____

Vocabulary Check **Write the vocabulary word that completes each sentence.**

41 A(n) _____ number is a number multiplied by itself, such as $4 \cdot 4$.

42 The _____ tells you how many times a base is multiplied by itself.

43 In the number 8^2, the _____ is 8.

44 Writing in Math The expression $(-4)^2$ has a value of 16 because $-4 \cdot (-4) = 16$. What is the value of the expression $(-6)^2$? Explain your reasoning.

_____

_____

STOP

Copyright © Glencoe/McGraw-Hill, a division of The McGraw-Hill Companies, Inc.

Square Roots

KEY Concept

The **square of a number** is the product of two identical factors. For example, 5^2 means $5 \cdot 5$. The product of $5 \cdot 5$ is 25.

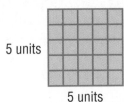

5 units

5 units

$5 \cdot 5 = 25$

Squaring a number and finding a **square root** are **inverse operations**. The square root of a number is one of its two equal factors.

The symbol for the positive square root is a **radical sign**, $\sqrt{}$.

The positive square root of 25 is written as $\sqrt{25}$. Since $5 \cdot 5 = 25$, we can rewrite $\sqrt{25}$ as $\sqrt{5 \cdot 5}$. So, 5 is the positive square root of 25.

$$\sqrt{25} = \sqrt{5 \cdot 5}$$
$$= 5$$

VOCABULARY

factor
a number that is multiplied by another number

inverse operations
operations which undo each other

radical sign
the symbol used to indicate a nonnegative square root, $\sqrt{}$

square of a number
the product of a number multiplied by itself; $4 \cdot 4$, or 4^2

square root
one of two equal factors of a number; if $a^2 = b$, then a is the square root of b

An area model represents a square number when the number of rows and the number of columns are equal. In the area model above there are 5 rows and 5 columns.

Example 1

Find the positive square root using an area model.

1. How many square units are shown? **16**

2. How many columns are shown? **4**

3. How many rows are shown? **4**

4. Are the number of rows and columns equal? **yes**

So, $\sqrt{16} = 4$.

GO ON

Copyright © Glencoe/McGraw-Hill, a division of The McGraw-Hill Companies, Inc.

Copyright © Glencoe/McGraw-Hill, a division of The McGraw-Hill Companies, Inc.

YOUR TURN!

Find the positive square root using an area model.

1. How many square units are shown? _____

2. How many columns are shown? _____

3. How many rows are shown? _____

4. Are the number of rows and columns equal? _____

So, $\sqrt{9} =$ _____.

Example 2

Find the positive square root of 49.

1. Write the expression.

$\sqrt{49}$

2. Name the factor pairs of 49.

$1 \cdot 49, \ 7 \cdot 7$

3. Replace 49 with the set of identical factors.

$\sqrt{7 \cdot 7}$

So, $\sqrt{49} = 7$.

YOUR TURN!

Find the positive square root of 64.

1. Write the expression.

_____

2. Name the factor pairs of 64.

_____

3. Replace 64 with the set of identical factors.

_____

So, _____.

Who is Correct?

Find the positive square root of 16.

Rondell
$\sqrt{16} = \sqrt{4 \cdot 4}$
$\sqrt{16} = 4$

Kenyon
$\sqrt{16} = \sqrt{8 \cdot 8}$
$\sqrt{16} = 8$

Orenda
$\sqrt{16} = \sqrt{1 \cdot 16}$
$\sqrt{16} = 16$

Circle the correct answer(s). Cross out incorrect answer(s).

Guided Practice

Find the positive square root using an area model.

1

$\sqrt{4} =$ _____

2 ☐

$\sqrt{1} =$ _____

Step by Step Practice

3 **Find the positive square root of 121.**

Step 1 Write the expression. _____

Step 2 Name the factor pairs of 121. _____

Step 3 Replace 121 with the set of identical factors. _____

Step 4 _____

Find the positive square root of each number.

4 169

Write the expression. _____

Name the factor pairs. _____

Replace 169 with the set of identical factors. _____

$\sqrt{169} =$ _____

5 36

Write the expression. _____

Name the factor pairs. _____

Replace 36 with the set of identical factors. _____

$\sqrt{36} =$ _____

6 144

Write the expression. _____

Name the factor pairs. _____

Replace 144 with the set of identical factors. _____

GO ON

$\sqrt{144} =$ _____

Copyright © Glencoe/McGraw-Hill, a division of The McGraw-Hill Companies, Inc.

Step by Step Problem-Solving Practice

Solve.

Problem-Solving Strategies
- ☐ Draw a diagram.
- ☐ Use logical reasoning.
- ☐ Solve a simpler problem.
- ☐ Work backward.
- ☑ Use a model.

7 **GARDENS** Mrs. Ramano's garden has an area of 144 square feet. The length of the garden and the width of the garden are the same. Use the positive square root of 144 to find the length of the garden.

Understand Read the problem. Write what you know.

The garden has an area of _____ square feet.

The length and the width are _____.

Plan Pick a strategy. One strategy is to use an area model.

Solve Use the square to create an area model. Divide the square into equal rows and columns until 144 square units are shown.

There are _____ rows and _____ columns.

$\sqrt{144} = \sqrt{\underline{} \cdot \underline{}} = \underline{}$

The length of the garden is _____ feet.

Check Use the inverse operation. _____ = 144

8 **TILES** Margarette is making a mosaic for her aunt's birthday present. She wants the mosaic to be square. She plans to use 81 tiles. Use the positive square root of 81 to determine how many rows and columns she will create.

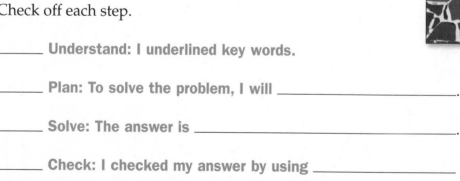

Check off each step.

_____ **Understand: I underlined key words.**

_____ **Plan: To solve the problem, I will** _____.

_____ **Solve: The answer is** _____.

_____ **Check: I checked my answer by using** _____

_____.

Copyright © Glencoe/McGraw-Hill, a division of The McGraw-Hill Companies, Inc.

9 **Reflect** Explain how knowing the factors of a number helps you find the number's positive square root.

_____

_____

_____

▶ Skills, Concepts, and Problem Solving

Find the positive square root using an area model.

10

$\sqrt{25} = $ _____

11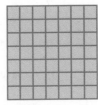

$\sqrt{49} = $ _____

Find the positive square root of each number.

12 81

Write the expression. _____

Name the factor pairs. _____

Replace 81 with the set of identical factors. _____

$\sqrt{81} = $ _____

13 100

Write the expression. _____

Name the factor pairs. _____

Replace 100 with the set of identical factors. _____

$\sqrt{100} = $ _____

14 196

Write the expression. _____

Name the factor pairs. _____

Replace 196 with the set of identical factors. _____

$\sqrt{196} = $ _____

GO ON

Copyright © Glencoe/McGraw-Hill, a division of The McGraw-Hill Companies, Inc.

Solve.

15 **CLASS SUPPLIES**   Mr. Orta wants to order a set of cubbie boxes with 169 small squares. The cubbie box has the same number of rows and columns. Use the positive square root of 169 to find the number of columns on the cubbie box.

_____

16 **CHESS**   A chessboard has 64 small squares that create a larger square. The number of rows is equal to the number of columns. Use the positive square root of 64 to find the number of rows on a chessboard.

_____

Vocabulary Check   **Write the vocabulary word that completes each sentence.**

17 Squaring a number and finding a square root are _____.

18 The _____ of a number is one of two equal factors of the number.

19 **Writing in Math**   Explain how an area model can help you find the positive square root of a number.

_____

_____

_____

▶ Spiral Review

Solve. (Lesson 6-1, p. 218)

20 **GAMES**   Mariska wants to arrange 8 rows of 8 game cards for her memory game. How many cards will she need? Write an expression using exponents to represent the number of cards. Find the value of the expression.

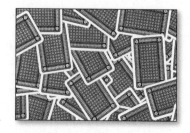

_____

Evaluate each expression. (Lesson 6-1, p. 218)

21 5^2 _____

22 7^2 _____

23 9^2 _____

24 6^2 _____

STOP

Copyright © Glencoe/McGraw-Hill, a division of The McGraw-Hill Companies, Inc.

Write an equation using exponents to represent each model.

1 _____

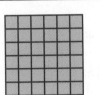

2 _____

Evaluate each expression.

3 1^2 _____

4 7^2 _____

5 9^2 _____

6 11^2 _____

Find the positive square root of each number.

7 64

Name the factor pairs. _____

Replace 64 with the set of identical factors. _____

$\sqrt{64} =$ _____

8 225

Name the factor pairs. _____

Replace 225 with the set of identical factors. _____

$\sqrt{225} =$ _____

9 169

Name the factor pairs. _____

Replace 169 with the set of identical factors. _____

$\sqrt{169} =$ _____

Solve.

10 **PUZZLES** Write a two-digit number the sum of whose digits is 13. The number is a squared number. _____

11 **CHAIRS** A small concert hall has 144 seats. The number of rows is equal to the number of columns. Use the positive square root of 144 to find the number of seats in each row.

Copyright © Glencoe/McGraw-Hill, a division of The McGraw-Hill Companies, Inc.

Approximate Square Roots

KEY Concept

If you know how to determine the square root of a positive whole number, you can **estimate** square roots.

The table below shows some common **square roots**.

Common Square Roots		
$\sqrt{1} = 1$	$\sqrt{36} = 6$	$\sqrt{121} = 11$
$\sqrt{4} = 2$	$\sqrt{49} = 7$	$\sqrt{144} = 12$
$\sqrt{9} = 3$	$\sqrt{64} = 8$	$\sqrt{169} = 13$
$\sqrt{16} = 4$	$\sqrt{81} = 9$	$\sqrt{196} = 14$
$\sqrt{25} = 5$	$\sqrt{100} = 10$	$\sqrt{225} = 15$

You can use these square roots to estimate other square roots.

You know that the number 5 is between 4 and 9. The number line below shows this relationship.

Since 5 is between 4 and 9, $\sqrt{5}$ must be between $\sqrt{4}$ and $\sqrt{9}$.

The number line above shows the relationship between the square roots.

$$\sqrt{4} = 2.00 \qquad \sqrt{5} \approx 2.23 \qquad \sqrt{9} = 3.00$$

This relationship can be shown as an inequality.

$$\sqrt{4} < \sqrt{5} < \sqrt{9}$$

VOCABULARY

estimate
a number close to an exact value; an estimate indicates *about* how much

square root
one of two equal factors of a number; if $a^2 = b$, then a is the square root of b

Memorizing the common square roots in the table will help you estimate square roots through 225.

Copyright © Glencoe/McGraw-Hill, a division of The McGraw-Hill Companies, Inc.

Example 1

Estimate $\sqrt{14}$ to the nearest whole number.

1. Write an inequality using common square roots.

$\sqrt{9} < \sqrt{14} < \sqrt{16}$

2. Find the values of the common square roots.

$\sqrt{9} = \sqrt{3 \cdot 3} = 3 \qquad \sqrt{16} = \sqrt{4 \cdot 4} = 4$

3. Plot the values of each square root on the number line.

So, $\sqrt{14}$ is between 3 and 4.

4. Since $\sqrt{14}$ is closer to $\sqrt{16}$ than $\sqrt{9}$, $\sqrt{14}$ is closer to the whole number 4.

YOUR TURN!

Estimate $\sqrt{8}$ to the nearest whole number.

1. Write an inequality using common square roots.

_____ < _____ < _____

2. Find the values of the common square roots.

$\sqrt{} = \sqrt{ \cdot } = \_\_ \qquad \sqrt{} = \sqrt{ \cdot } = \_\_$

3. Plot the values of each square root on the number line.

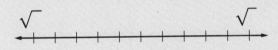

So, $\sqrt{8}$ is between _____ and _____.

4. Since $\sqrt{8}$ is closer to $\sqrt{}$ than $\sqrt{}$, $\sqrt{8}$ is closer to the whole

number _____.

Copyright © Glencoe/McGraw-Hill, a division of The McGraw-Hill Companies, Inc.

GO ON

Example 2

Choose a reasonable estimate for $\sqrt{109}$.

1. Write an inequality using common square roots.

$\sqrt{100} < \sqrt{109} < \sqrt{121}$

2. Find the values of the common square roots.

$\sqrt{100} = \sqrt{10 \cdot 10} = 10$ $\sqrt{121} = \sqrt{11 \cdot 11} = 11$

So, $\sqrt{109}$ is between 10 and 11.

4. Since $\sqrt{109}$ is closer to $\sqrt{100}$ than $\sqrt{121}$, $\sqrt{109}$ is closer to the whole number 10.

5. Circle the reasonable estimate.

(10.4)   10.8   11.2

YOUR TURN!

Choose a reasonable estimate for $\sqrt{50}$.

1. Write an inequality using common square roots.

$\sqrt{\underline{}} < \sqrt{50} < \sqrt{\underline{}}$

2. Find the values of the common square roots.

$\sqrt{\underline{}} = \sqrt{\underline{}} = \underline{}$ $\sqrt{\underline{}} = \sqrt{\underline{}} = \underline{}$

So, $\sqrt{50}$ is between ___ and ___.

4. Since $\sqrt{50}$ is closer to $\sqrt{\underline{}}$ than $\sqrt{\underline{}}$, $\sqrt{50}$ is closer to the whole number ___.

5. Circle the reasonable estimate.

6.8   7.1   7.6

Who is Correct?

Estimate $\sqrt{18}$ to the nearest whole number.

Connor
$\sqrt{16} < \sqrt{18} < \sqrt{25}$
Falls between 4 and 5; it is closer to 4.

Rachelle
$\sqrt{16} < \sqrt{18} < \sqrt{20}$
Falls between 9 and 10; it is closer to 9.

Masao
$\sqrt{16} < \sqrt{18} < \sqrt{25}$
Falls between 4 and 5; it is closer to 5.

Circle correct answer(s). Cross out incorrect answer(s).

Copyright © Glencoe/McGraw-Hill, a division of The McGraw-Hill Companies, Inc.

Copyright © Glencoe/McGraw-Hill, a division of The McGraw-Hill Companies, Inc.

 Guided Practice

Write an inequality using common square roots.

1 $\sqrt{} < \sqrt{84} < \sqrt{}$

2 $\sqrt{} < \sqrt{136} < \sqrt{}$

3 $\sqrt{} < \sqrt{21} < \sqrt{}$

4 $\sqrt{} < \sqrt{47} < \sqrt{}$

5 $\sqrt{} < \sqrt{13} < \sqrt{}$

6 $\sqrt{} < \sqrt{200} < \sqrt{}$

Step by **Step Practice**

7 Estimate $\sqrt{92}$ to the nearest whole number.

Step 1 Write an inequality using common square roots.

$\sqrt{} < \sqrt{92} < \sqrt{}$

Step 2 Find the values of the common square roots.

$\sqrt{} = \sqrt{} = \underline{}$ $\sqrt{} = \sqrt{} = \underline{}$

Step 3 Plot the values of each square root on the number line.

So, $\sqrt{92}$ is between ____ and ____.

Step 4 Since $\sqrt{92}$ is closer to $\sqrt{}$ than $\sqrt{}$, $\sqrt{92}$ is closer to the whole number _____.

Estimate each square root to the nearest whole number. Plot each value on a number line.

8 $\sqrt{7}$ is close to the whole number _____.

9 $\sqrt{68}$ is close to the whole number _____.

GO ON

Choose a reasonable estimate for each square root.

10 $\sqrt{22}$

$$\sqrt{\rule{1em}{0.4pt}} < \sqrt{22} < \sqrt{\rule{1em}{0.4pt}}$$

$$\sqrt{\rule{1em}{0.4pt}} = \rule{1em}{0.4pt} \qquad \sqrt{\rule{1em}{0.4pt}} = \rule{1em}{0.4pt}$$

4.2   4.7   5.1

11 $\sqrt{139}$

$$\sqrt{\rule{1em}{0.4pt}} < \sqrt{139} < \sqrt{\rule{1em}{0.4pt}}$$

$$\sqrt{\rule{1em}{0.4pt}} = \rule{1em}{0.4pt} \qquad \sqrt{\rule{1em}{0.4pt}} = \rule{1em}{0.4pt}$$

11.8   12.1   12.6

Step by Step Problem-Solving Practice

Problem-Solving Strategies
☐ Draw a diagram.
☑ Use logical reasoning.
☐ Work backward.
☐ Solve a simpler problem.
☐ Look for a pattern.

Solve.

12 **PATIOS**   Troy has a square patio in his backyard. The patio has an area of 172 square feet. Estimate the length and width of the patio.

Understand   Read the problem. Write what you know.

The patio is the shape of a _____.

The length and width of squares are _____.

The total area is _____ square feet.

Plan   Pick a strategy. One strategy is to use logical reasoning.

Solve   Find the common square roots that are close in value.

$$\sqrt{\rule{1em}{0.4pt}} < \sqrt{172} < \sqrt{\rule{1em}{0.4pt}}$$

So, $\sqrt{172}$ is between _____ and _____.

Since $\sqrt{172}$ is closer to $\sqrt{\rule{1em}{0.4pt}}$ than $\sqrt{\rule{1em}{0.4pt}}$, $\sqrt{172}$ is closer to the whole number _____.

Check   Use a number line to check your answer.

Copyright © Glencoe/McGraw-Hill, a division of The McGraw-Hill Companies, Inc.

13 **GARDENS** Marlene is creating a garden in her yard. The garden will cover an area of 15 square feet. The garden will be arranged as a square. Estimate the length and width of the garden to the nearest whole number.
Check off each step.

_____ **Understand: I underlined key words.**

_____ **Plan: To solve the problem, I will** _____.

_____ **Solve: The answer is** _____

_____.

_____ **Check: I checked my answer by using** _____.

14 **Reflect** How is it helpful to estimate?

_____

_____

▶ Skills, Concepts, and Problem Solving

Write an inequality using common square roots.

15 $\sqrt{\_\_} < \sqrt{5} < \sqrt{\_\_}$

16 $\sqrt{\_\_} < \sqrt{29} < \sqrt{\_\_}$

17 $\sqrt{\_\_} < \sqrt{217} < \sqrt{\_\_}$

18 $\sqrt{\_\_} < \sqrt{191} < \sqrt{\_\_}$

Estimate each square root to the nearest whole number. Plot each value on a number line.

19 $\sqrt{11}$ is close to the whole number _____.

20 $\sqrt{143}$ is close to the whole number _____.

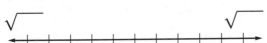

21 $\sqrt{38}$ is close to the whole number _____.

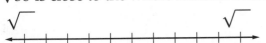

Copyright © Glencoe/McGraw-Hill, a division of The McGraw-Hill Companies, Inc.

Choose a reasonable estimate for each square root.

22 $\sqrt{88}$

    8.8   9.1   9.4

23 $\sqrt{57}$

    5.7   7.5   7.9

24 $\sqrt{71}$

    8.4   8.7   9.2

25 $\sqrt{63}$

    6.3   7.4   7.9

26 $\sqrt{14}$

    3.2   3.7   4.1

27 $\sqrt{2}$

    1.1   1.4   1.9

Solve.

28 **GAMES** A square game board had a total of 64 one-inch squares and a $\frac{1}{4}$-inch border. Estimate the length and width of the game board.

_____

29 **AREA** Mr. Fox's bedroom is shaped like a square. The area is 120 square feet. Estimate the length and width of the room.

_____

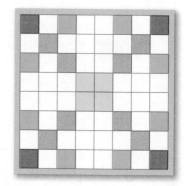

Vocabulary Check **Write the vocabulary word that completes each sentence.**

30 A(n) _____ indicates about how much.

31 The _____ of 169 is 13.

32 **Writing in Math** Explain how to find an estimate using a number line.

_____

_____

 Spiral Review (Lesson 6-2, p. 225)

33 **VIDEO GAMES** Orlando is playing a dance video game. The game pad is made of 9 stepping squares. The number of rows and columns are the same. How many columns of stepping squares are on the game pad? Use the positive square root of 9 to determine the number of columns.

_____

Copyright © Glencoe/McGraw-Hill, a division of The McGraw-Hill Companies, Inc.

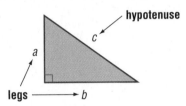

Pythagorean Theorem

KEY Concept

You know that a triangle has three sides. A right triangle is a special triangle that has one right angle and two acute angles.

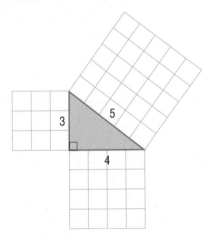

The sides of a right triangle have special names. A right triangle has two sides called **legs**. The legs in the triangle above are labeled a and b. The third side of the triangle is called the **hypotenuse**. It is labeled c.

The **Pythagorean Theorem** states that $a^2 + b^2 = c^2$. To prove the Pythagorean Theorem, count the number of squares beside each side of the triangle above.

You can also use the formula.

$a^2 + b^2 = c^2$	Use the formula.
$3^2 + 4^2 = 5^2$	Substitute the measures of each side.
$(3 \cdot 3) + (4 \cdot 4) = (5 \cdot 5)$	Simplify the exponents.
$9 + 16 = 25$	Multiply.
$25 = 25$	Add.

VOCABULARY

hypotenuse
the side opposite the right angle in a right triangle

legs
the two sides of a right triangle that form the right angle

Pythagorean Theorem
in a right triangle, the square of the length of the hypotenuse, c, is equal to the sum of the squares of the lengths of the legs, a and b
$c^2 = a^2 + b^2$

square of a number
the product of a number multiplied by itself; $4 \cdot 4$, or 4^2

square root
one of two equal factors of a number; if $a^2 = b$, then a is the square root of b

Given the measures of any two sides of a right triangle, you can use the Pythagorean Theorem to find the unknown length of the third side.

Copyright © Glencoe/McGraw-Hill, a division of The McGraw-Hill Companies, Inc.

Example 1

Find the length of the hypotenuse of the right triangle.

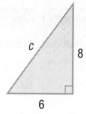

1. From the figure, you know $a = 6$ and $b = 8$.

2. Substitute the values of a and b in the Pythagorean Theorem and solve for c.

$a^2 + b^2 = c^2$	Use the formula.
$6^2 + 8^2 = c^2$	Substitute the measure of each side.
$36 + 64 = c^2$	Simplify the exponents.
$100 = c^2$	Add 36 and 64.
$\sqrt{100} = \sqrt{c^2}$	Take the square root of both sides.
$10 = c$	Simplify the square root.

3. The length of the hypotenuse is 10 units.

YOUR TURN!

Find the length of the hypotenuse of the right triangle.

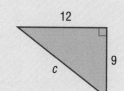

1. From the figure, you know

   $a =$ _____ and $b =$ _____.

2. Substitute the values of a and b in the Pythagorean Theorem and solve for c.

   $$a^2 + b^2 = c^2$$

   $$\underline{\qquad}^2 + \underline{\qquad}^2 = c^2$$

   $$\underline{\qquad} + \underline{\qquad} = c^2$$

   $$\underline{\qquad} = c^2$$

   $$\underline{\qquad} = \sqrt{c^2}$$

   $$\underline{\qquad} = c$$

3. The length of the hypotenuse is _____ units.

Copyright © Glencoe/McGraw-Hill, a division of The McGraw-Hill Companies, Inc.

Example 2

Find the length of the leg of the right triangle.

1. From the figure, you know $b = 60$ and $c = 75$.

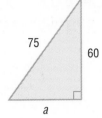

2. Substitute the values of b and c in the Pythagorean Theorem and solve for a.

$a^2 + b^2 = c^2$	Use the formula.
$a^2 + 60^2 = 75^2$	Substitute the measure of each side.
$a^2 + 3{,}600 = 5{,}625$	Simplify the exponents.
$a^2 = 5{,}625 - 3{,}600$	Subtract 3,600 from 5,625.
$a^2 = 2{,}025$	Simplify.
$\sqrt{a^2} = \sqrt{2{,}025}$	Take the square root of both sides.
$a = 45$	Simplify the square root.

3. The length of the leg is 45 units.

YOUR TURN!

Find the length of the leg of the right triangle.

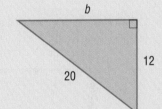

1. From the figure, you know

$a = $ _____ and $c = $ _____.

2. Substitute the values of a and c in the Pythagorean Theorem and solve for b.

$$a^2 + b^2 = c^2$$

_____ $+ b^2 = $ _____

_____ $+ b^2 = $ _____

$b^2 = $ _____

$b^2 = $ _____

$\sqrt{b^2} = $ _____

$b = $ _____

3. The length of the leg is _____ units.

You can also do this in reverse. If the sum of the squares of the two smaller sides of a given triangle equals the square of the larger side, then the triangle is a right triangle.

Copyright © Glencoe/McGraw-Hill, a division of The McGraw-Hill Companies, Inc.

Example 3

Determine if the triangle is a right triangle, using the Pythagorean Theorem.

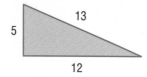

1. Determine which side is the longest. **13**

2. Substitute the values into $a^2 + b^2 = c^2$, ensuring that the largest number is c, and solve.

$a^2 + b^2 = a^2$	Use the formula.
$5^2 + 12^2 = 13^2$	Substitute the measure of each side.
$25 + 144 = 169$	Simplify the exponents.
$169 = 169$ ✓	Add 25 and 144.

3. Since this is a true statement, the triangle is a right triangle.

YOUR TURN!

Determine if the triangle is a right triangle, using the Pythagorean Theorem.

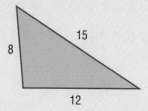

1. Determine which side is the longest. _____

2. Substitute the values into $a^2 + b^2 = c^2$, ensuring that the largest number is c, and solve.

$$a^2 + b^2 = c^2$$

_____ + _____ = _____

_____ + _____ = _____

_____ ◯ _____

3. Since this is a _____ statement, the triangle _____ a right triangle.

Who is Correct?

Mr. Fernandez told his students to draw and label the sides of a right triangle. Which student's triangle is correct?

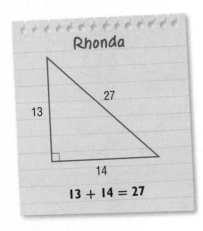

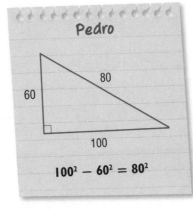

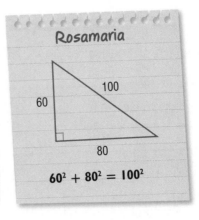

Circle correct answer(s). Cross out incorrect answer(s).

Copyright © Glencoe/McGraw-Hill, a division of The McGraw-Hill Companies, Inc.

▶ Guided Practice

Find the length of the hypotenuse of the right triangle.

1

$c =$ _____

2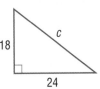

$c =$ _____

Step by Step Practice

3 Find the length of the leg of the right triangle.

Step 1 Find the lengths of sides a and c.

$a =$ _____ and $c =$ _____

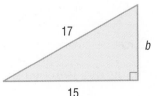

Step 2 Substitute the values of a and c in the Pythagorean Theorem and solve for b.

$$a^2 + b^2 = c^2$$
$$\underline{}^2 + b^2 = \underline{}^2$$
$$\underline{} + b^2 = \underline{}$$
$$b^2 = \underline{}$$
$$\sqrt{b^2} = \underline{}$$
$$b = \underline{}$$

The length of the leg is _____ units.

Find the length of the leg of each right triangle.

4
$$a^2 + b^2 = c^2$$
$$a^2 + \underline{}^2 = \underline{}^2$$
$$a^2 + \underline{} = \underline{}$$
$$a^2 = \underline{}$$
$$\sqrt{a^2} = \underline{}$$
$$a = \underline{} \text{ units}$$

5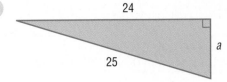

$a =$ _____ units

GO ON

Copyright © Glencoe/McGraw-Hill, a division of The McGraw-Hill Companies, Inc.

Determine if each triangle is a right triangle, using the Pythagorean Theorem.

6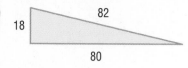

7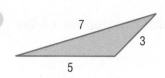

_____   _____

Step by Step Problem-Solving Practice

Solve.

Problem-Solving Strategies
☑ Draw a diagram.
☐ Look for a pattern.
☐ Act it out.
☐ Solve a simpler problem.
☐ Work backward.

8 **HOME IMPROVEMENT**   Jerry's ladder is resting against a wall. The top of the ladder touches the wall at a height of 12 feet. The bottom of his ladder is 9 feet away from the base of the wall. How long is Jerry's ladder?

Understand   Read the problem. Write what you know.

The ladder touches the wall at a height of _____ feet.

The ladder is _____ feet away from the base of the wall.

Plan   Pick a strategy. One strategy is to draw a diagram. Then, use the Pythagorean Theorem to find the length of the ladder.

Solve   Substitute 9 for a and 12 for b in the Pythagorean Theorem and solve for c.

$$a^2 + b^2 = c^2$$
$$\underline{}^2 + \underline{}^2 = c^2$$
$$\underline{} + \underline{} = c^2$$
$$\underline{} = c^2$$
$$\underline{} = \sqrt{c^2}$$
$$\underline{} = c$$

The length of the ladder is _____ feet.

Check   Use a calculator to check your answer.

Copyright © Glencoe/McGraw-Hill, a division of The McGraw-Hill Companies, Inc.

9 CONSTRUCTION A staircase to the attic in Cassandra's house has a length of 10 feet. The top of the stairs meets the wall at a height of 8 feet. How far away is the bottom of the staircase from the wall? Check off each step.

_____ **Understand: I underlined key words.**

_____ **Plan: To solve the problem, I will** _____.

_____ **Solve: The answer is** _____.

_____ **Check: I checked my answer by** _____.

10 ADVERTISING Splash Village advertised a water slide that is 20 yards long and shoots straight down to 12 yards from the base of the steps to the end of the slide. What is the height of the steps that reach the top of the slide?

_____

11 **Reflect** Can line segments with lengths 30 inches, 30 inches, and 50 inches form a right triangle? Explain.

_____

_____

▶ Skills, Concepts, and Problem Solving

Find the length of the leg or hypotenuse of each right triangle.

12

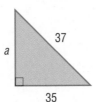

$a =$ _____ units

13

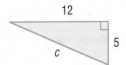

$c =$ _____ units

Determine if each triangle is a right triangle, using the Pythagorean Theorem.

14

_____

15

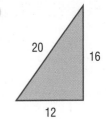

_____

GO ON

Copyright © Glencoe/McGraw-Hill, a division of The McGraw-Hill Companies, Inc.

Solve.

16 **TRAVEL** Reggie left his house and drove 20 miles due east and then 15 miles due south. If Reggie follows a straight line to his house, how far is Reggie from his house?

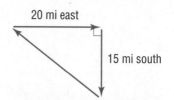

_____

17 **DISTANCE** A lighthouse on a cliff near Gina's house is 12 meters tall. Gina stood 9 meters away from the base of the lighthouse. How far away was Gina from the top of the lighthouse?

_____

Vocabulary Check **Write the vocabulary word that completes each sentence.**

18 A(n) _____ is the side opposite the right angle in a right triangle.

19 The _____ states that the sum of the squares of the lengths of the legs in a right triangle is equal to the square of the length of the hypotenuse.

20 **Writing in Math** Explain how to find the length of the hypotenuse of a right triangle when the lengths of the legs are known.

_____

_____

▶ **Spiral Review**

Choose a reasonable estimate for each square root. (Lesson 6-3, p. 232)

21 $\sqrt{104}$

   10.2   10.4   11.1

22 $\sqrt{46}$

   4.6   6.2   6.8

23 $\sqrt{79}$

   7.9   8.2   8.9

24 $\sqrt{128}$

   11.0   11.3   12.1

STOP

Copyright © Glencoe/McGraw-Hill, a division of The McGraw-Hill Companies, Inc.

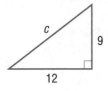
Find the length of the leg or hypotenuse of each right triangle.

1

c
9
12

$c =$ _____ units

2

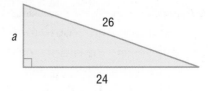

26
a
24

$a =$ _____ units

Estimate each square root to the nearest whole number. Plot each value on a number line.

3 $\sqrt{15}$ is close to the whole number _____.

4 $\sqrt{165}$ is close to the whole number _____.

Choose a reasonable estimate for each square root.

5 $\sqrt{8}$

2.1   2.5   2.8

6 $\sqrt{78}$

8.8   9.1   9.5

Solve.

7 **FLAGS**   A flagpole stands in front of a city building. The base of the flagpole is 8 yards away from a spotlight that shines on the flag. The distance from the spotlight to the top of the flagpole is 10 yards. How tall is the flagpole?

_____

8 **RECTANGLES**   What is the length of the diagonal of rectangle *MNOP*?

_____

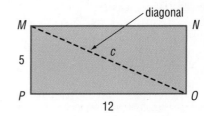

diagonal
M   N
5   c
P   O
12

Copyright © Glencoe/McGraw-Hill, a division of The McGraw-Hill Companies, Inc.

Introduction to Slope

KEY Concept

When describing a line on a **coordinate grid**, you can describe the "steepness," or **slope**, of the line. Lines that move upward and to the right have a positive slope. Lines that move downward and to the left have a negative slope.

$$\text{slope} = \frac{\text{number of units up }(+)\text{ or down }(-)}{\text{number of units right }(+)\text{ or left }(-)} = \frac{\text{rise}}{\text{run}}$$

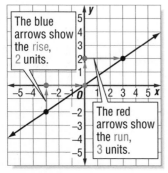

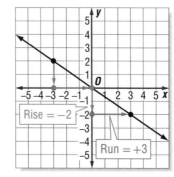

$$\frac{\text{rise}}{\text{run}} = \frac{+2}{+3} = \text{positive slope} \qquad \frac{\text{rise}}{\text{run}} = \frac{-2}{+3} = \text{negative slope}$$

Look at each graph from left to right in order to determine if the slope is positive or negative. If the line goes up, the slope is positive. If the line goes down, the slope is negative.

VOCABULARY

coordinate grid
a grid in which a horizontal number line (*x*-axis) and a vertical number line (*y*-axis) intersect at their zero points

origin
the point of intersection of the *x*-axis and the *y*-axis in a coordinate system

slope
the rate of change between any two points on a line; the ratio of vertical change to horizontal change

x-coordinate
the first number of an ordered pair

y-coordinate
the second number of an ordered pair

A slope is negative if the value of the rise or the value of the run is negative.

Example 1

Find the slope of the line.

1. The rise is $+4$ units.

2. The run is $+5$ units.

3. The slope is positive.

4. $\dfrac{\text{rise}}{\text{run}} = \dfrac{+4}{+5} = \dfrac{4}{5}$

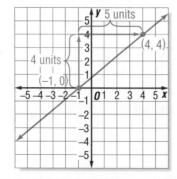

Copyright © Glencoe/McGraw-Hill, a division of The McGraw-Hill Companies, Inc.

Copyright © Glencoe/McGraw-Hill, a division of The McGraw-Hill Companies, Inc.

YOUR TURN!

Find the slope of the line.

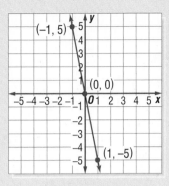

1. The rise is _____ units.

2. The run is _____ units.

3. The slope is _____.

4. $\dfrac{\text{rise}}{\text{run}} = \dfrac{\boxed{}}{\boxed{}} = $ _____

Example 2

Graph another point on the line, given the point (3, 4) and the slope $\dfrac{1}{4}$.

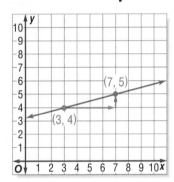

1. Find the point $(3, 4)$.

2. The "run" is $+4$. Add 4 to the x-coordinate.

   $3 + 4 = 7$
   $x = 7$

3. The "rise" is $+1$. Add 1 to the y-coordinate.

   $4 + 1 = 5$
   $y = 5$

4. Name and graph the new point.
   $(x, y) = (7, 5)$

YOUR TURN!

Graph another point on the line, given the point (5, 4) and the slope $-\dfrac{1}{3}$.

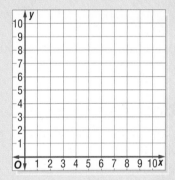

1. Find the point (_____, _____).

2. The "run" is _____. _____
   the x-coordinate.

   5 _____ = _____

   $x = $ _____

3. The "rise" is _____. _____
   the y-coordinate.

   4 _____ = _____

   $y = $ _____

4. Name and graph the new point.

   $(x, y) = $ (_____, _____)

Who is Correct?

Find the slope of the line.

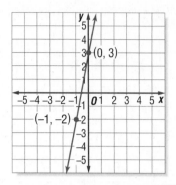

Sherita

$$\text{Slope} = \frac{\text{rise}}{\text{run}} = \frac{5}{1} = 5$$

Miguel

$$\text{Slope} = \frac{\text{rise}}{\text{run}} = \frac{-5}{1} = -5$$

Candida

$$\text{Slope} = \frac{\text{rise}}{\text{run}} = \frac{-1}{5} = \frac{1}{5}$$

Circle the correct answer(s). Cross out incorrect answer(s).

▶ Guided Practice

Find the slope of each line.

1 The rise is _____ units.

The run is _____ units.

The slope is _____.

$$\frac{\text{rise}}{\text{run}} = \frac{\boxed{}}{\boxed{}} = \text{\_\_\_\_\_} = \text{\_\_\_\_\_}$$

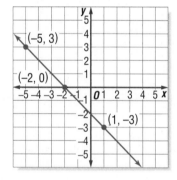

2

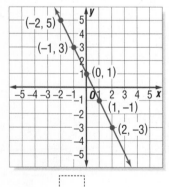

$$\frac{\text{rise}}{\text{run}} = \frac{\boxed{}}{\boxed{}} = \text{\_\_\_\_\_}$$

3

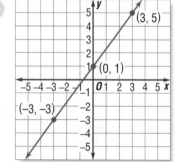

$$\frac{\text{rise}}{\text{run}} = \frac{\boxed{}}{\boxed{}} = \text{\_\_\_\_\_}$$

Copyright © Glencoe/McGraw-Hill, a division of The McGraw-Hill Companies, Inc.

4 Graph another point on the line, given the point (8, 1) and the slope −3.

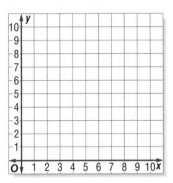

Hint: You can either consider the "rise" to be negative or the "run" to be negative.

Step 1 Find the point (8, 1).

Step 2 The "run" is _____.

8 _____ = _____ x = _____

Step 3 The "rise" is _____.

1 _____ = _____ y = _____

Step 4 Name and graph the new point.

(x, y) = (____, ____)

Graph another point on each line, given one point on the line and the slope.

5

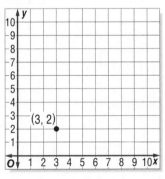

The slope is $\frac{1}{2}$.

6

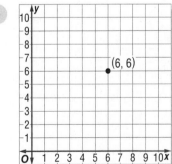

The slope is $-\frac{3}{4}$.

GO ON

Copyright © Glencoe/McGraw-Hill, a division of The McGraw-Hill Companies, Inc.

Step by Step Problem-Solving Practice

Solve.

Problem-Solving Strategies
☐ Draw a diagram.
☐ Use logical reasoning.
☑ Use a formula.
☐ Solve a simpler problem.
☐ Work backward.

7 RAMPS Laura is learning how to do turns on a skateboard ramp. The ramp is 4 feet long and 2 feet tall. What is the slope of the ramp?

Understand Read the problem. Write what you know.

The ramp length is _____.

The ramp height is _____.

Plan Pick a strategy. One strategy is to use a formula. Use the formula for slope to solve the problem.

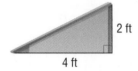
2 ft
4 ft

Solve Find the rise and run and then simplify.

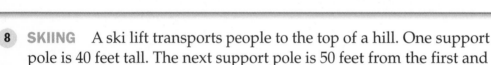

$$\text{slope} = \frac{\square}{\square} \div \frac{\square}{\square} = \frac{\square}{\square}$$

Check Use a coordinate grid. Start at the origin and move 4 units to the right and 2 units upward. Graph that point and draw a line segment to connect the points. Use these two points to find the slope of the line segment.

8 SKIING A ski lift transports people to the top of a hill. One support pole is 40 feet tall. The next support pole is 50 feet from the first and is 100 feet tall. Find the slope of the wire between the poles. Check off each step.

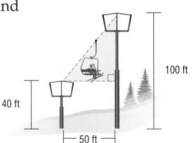
100 ft
40 ft
50 ft

_____ Understand: I underlined key words.

_____ Plan: To solve this problem, I will _____.

_____ Solve: The answer is _____.

_____ Check: I checked my answer by _____

_____.

9 Reflect How does viewing the line on the graph before looking at numeric values help you check your answers?

_____

_____

Copyright © Glencoe/McGraw-Hill, a division of The McGraw-Hill Companies, Inc.

Copyright © Glencoe/McGraw-Hill, a division of The McGraw-Hill Companies, Inc.

 Skills, Concepts, and Problem Solving

Find the slope of each line.

10 The rise is _____ units.

The run is _____ units.

The slope is _____.

$\dfrac{\text{rise}}{\text{run}} = \dfrac{\boxed{}}{\boxed{}} = \underline{}$

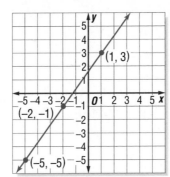

11

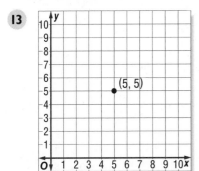

$\dfrac{\text{rise}}{\text{run}} = \dfrac{\boxed{}}{\boxed{}} = \underline{}$

12

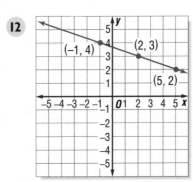

$\dfrac{\text{rise}}{\text{run}} = \dfrac{\boxed{}}{\boxed{}} = \underline{}$

Graph another point on each line, given one point on the line and the slope.

13

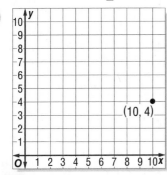

The slope is $-\dfrac{3}{2}$.

14

The slope is $\dfrac{1}{2}$.

15

The slope is $-\dfrac{1}{5}$.

16

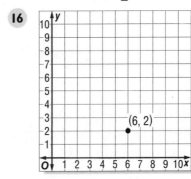

The slope is $-\dfrac{1}{2}$.

GO ON

Solve.

17 **SAILS** A sail on a sailboat is 24 feet tall and 12 feet across the bottom. What is the slope of the sail?

_____

18 **LADDERS** A ladder is leaning against a wall. It touches the wall 60 feet from the ground. The bottom of the ladder is 20 feet from the wall. What is the slope of the ladder?

_____

Vocabulary Check **Write the vocabulary word that completes each sentence.**

19 The _____ of a line refers to the "steepness" of the line.

20 The _____ is the place where the *x*-axis and *y*-axis cross.

21 **Writing in Math** Explain what happens to the steepness of the line when the value of the slope increases.

_____

_____

▶ **Spiral Review**

Determine if each triangle is a right triangle using the Pythagorean Theorem. (Lesson 6-4, p. 239)

22

18 24 21

23

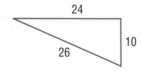

24 26 10

_____   _____

Solve. (Lesson 6-4, p. 239)

24 **SHADOWS** The shadow of the locust tree in Quanah's backyard is 5 feet long. The distance from the end of the shadow to the top of the tree is 13 feet. How tall is the tree?

_____

STOP

Copyright © Glencoe/McGraw-Hill, a division of The McGraw-Hill Companies, Inc.

Slope Formula

KEY Concept

The **slope** of a line illustrates the ratio of the number of units of rise to the number of units of run for a **linear function**. You can find the slope of a line from the graph of that line or by using the slope formula.

$$\text{slope} = m = \frac{\Delta y}{\Delta x} = \frac{\text{change in } y}{\text{change in } x}$$

> A lowercase m represents slope.

$$m = \frac{y_2 - y_1}{x_2 - x_1}, \text{ where } x_2 \neq x_1$$

Look at the graph at the right. Choose two points on the line, such as $(-3, -2)$ and $(0, 0)$.

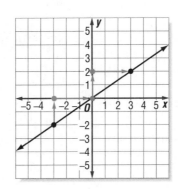

Find the slope by substituting the points $(0, 0)$ and $(3, 2)$ in the formula.

$$m = \frac{y_2 - y_2}{x_2 - x_2} = \frac{0 - 2}{0 - 3} = \frac{-2}{-3} = \frac{2}{3}$$

> Remember that a negative number, divided by a negative number equals a positive quotient.

You can continue to move up 2 units and right 3 units because the slope of the line is constant.

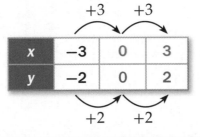

VOCABULARY

slope
the rate of change between any two points on a line; the ratio of vertical change to horizontal change

linear function
a function whose graph is a straight line

In the slope formula, the change in y-values is the numerator, and the change in x-values is the denominator.

Copyright © Glencoe/McGraw-Hill, a division of The McGraw-Hill Companies, Inc.

Example 1

Determine the slope of the graph.

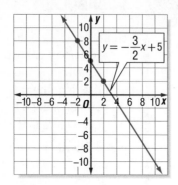

1. Complete a function table for the graph.

Point	A	B	C
x	2	0	−2
y	2	5	8

2. Substitute the *x* and *y* values in the slope formula to find the slope of the line.

$$m = \frac{y_2 - y_1}{x_2 - x_1} = \frac{5 - 2}{0 - 2}$$

$$= -\frac{3}{2}$$

3. The slope of the line is $-\frac{3}{2}$.

YOUR TURN!

Determine the slope of the graph.

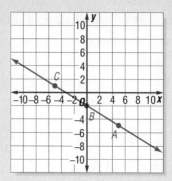

1. Complete a function table for the graph.

Point	A	B	C
x			
y			

2. Substitute the *x* and *y* values in the slope formula to find the slope of the line.

$$m = \frac{y_2 - y_1}{x_2 - x_1} = \frac{\boxed{} - \boxed{}}{\boxed{} - \boxed{}}$$

$$= \underline{}$$

3. The slope of the line is _____.

Copyright © Glencoe/McGraw-Hill, a division of The McGraw-Hill Companies, Inc.

Example 2

Graph y = 2x + 3 and determine the slope of the line.

1. Complete a function table for the equation.

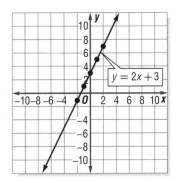

x	−2	−1	0	1	2
y	−1	1	3	5	7

2. Graph the ordered pairs and draw the line.

$y = 2x + 3$

3. Using two points on the graph, determine the slope.

$$m = \frac{1 - (-1)}{-1 - (-2)} = \frac{1 + 1}{-1 + 2} = \frac{2}{1} = 2$$

YOUR TURN!

Graph $y = \frac{1}{2}x - 8$ and determine the slope of the line.

1. Complete a function table for the equation.

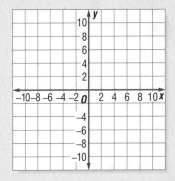

x	−2	0	2	4	6
y					

2. Graph the ordered pairs and draw the line.

3. Using two points on the graph, determine the slope.

$$m = \frac{\Box - \Box}{\Box - \Box} = \frac{\Box}{\Box} = \underline{}$$

Who is Correct?

What is the slope of the line on the graph?

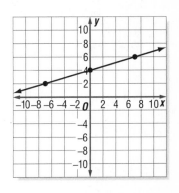

Dale

Slope $= \dfrac{0 - 7}{4 - 6}$

$= \dfrac{-7}{-2} = \dfrac{7}{2}$

Landon

Slope $= \dfrac{4 - 6}{0 - 7}$

$= \dfrac{-2}{-7} = \dfrac{2}{7}$

Jena

Slope $= \dfrac{4 - 2}{-7 - 0}$

$= \dfrac{2}{-7} = \dfrac{-2}{7}$

Circle correct answer(s). Cross out incorrect answer(s).

Copyright © Glencoe/McGraw-Hill, a division of The McGraw-Hill Companies, Inc.

Guided Practice

Use the graph to answer each question.

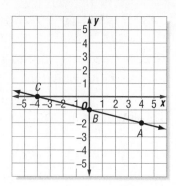

1 What is the location of Point A? _____

2 What is the location of Point B? _____

Step by Step Practice

3 Determine the slope of the graph.

Step 1 Complete a function table for the graph.

x	0	1	2	3
y				

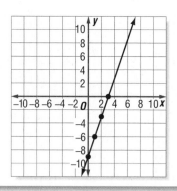

Step 2 Substitute the x and y values in the slope formula to find the slope of the line.

$m = \dfrac{\boxed{} - \boxed{}}{\boxed{} - \boxed{}} = \dfrac{\boxed{}}{\boxed{}} = \dfrac{\boxed{}}{\boxed{}}$ or _____

Step 3 The slope of the line is _____.

Graph each equation and determine its slope.

4 $y = -\dfrac{3}{4}x - 3$

x	−4	0	4
y			

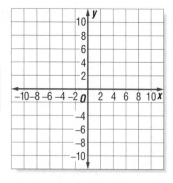

5 $y = 6x + 2$

x	−1	0	1
y			

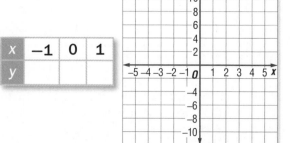

The slope is _____.

The slope is _____.

Copyright © Glencoe/McGraw-Hill, a division of The McGraw-Hill Companies, Inc.

Step by Step Problem-Solving Practice

Solve.

6 **PRICES** Purified water costs $2 per gallon. Graph an equation to represent the cost of purchasing x gallons of water.

Copyright © Glencoe/McGraw-Hill, a division of The McGraw-Hill Companies, Inc.

Understand   Read the problem. Write what you know.

Purifed water costs _____.

Plan   Pick a strategy. One strategy is to write an equation.

Let x represent the number of gallons and y represent the cost.

_____

Solve   Use the equation to make a function table.

Number of Gallons, x	0	1	2	3
Cost, y				

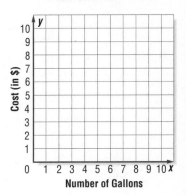

Purified Water Prices

Graph the ordered pairs from the table and draw the line on the graph.

Use the formula for slope.

$$m = \frac{y_2 - y_1}{x_2 - x_1} = \frac{\boxed{} - \boxed{}}{\boxed{} - \boxed{}} = \underline{\qquad}$$

Check $slope = \dfrac{rise}{run} = \dfrac{\boxed{}}{\boxed{}}$ or _____

To find the slope, move up _____ unit(s) and to the right _____ unit(s).

GO ON

Problem-Solving Strategies

☐ Draw a diagram.
☐ Look for a pattern.
☐ Guess and check.
☑ Write an equation.
☐ Work backward.

7 **MONEY** Julián earns $4 for every car he washes. Graph an equation to represent the total amount Julián earns if he washes x cars. Include at least three points on the graph.

Check off each step.

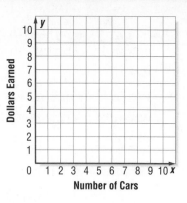

_____ **Understand: I underlined key words.**

_____ **Plan: To solve the problem I will** _____.

_____ **Solve: The equation is** _____. **The slope is** _____.

_____ **Check: I checked my answer by** _____.

8 **Reflect** Draw two lines on the graph to the right. One line should have a slope of $\frac{1}{3}$. The second line should have a slope of $-\frac{1}{3}$. Describe the direction of the two lines.

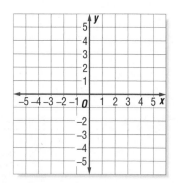

_____

_____

_____

_____

 ## Skills, Concepts, and Problem Solving

Graph each equation and determine its slope.

9 $y = -6x + 4$

x	−1	0	1	2
y				

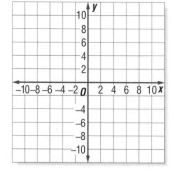

The slope is _____.

Copyright © Glencoe/McGraw-Hill, a division of The McGraw-Hill Companies, Inc.

10 **CANOEING** Ines can row a canoe 3 miles in 1 hour. Graph an equation to represent the number of miles Ines can row a canoe in x hours.

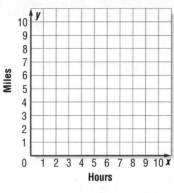

Write an equation. Let x represent the hours that Ines rows a canoe and y represent the number of miles.

_____

The slope is _____.

11 **BOOK CLUB** The Rosemill Book Club reads 2 books each month. Graph an equation to represent the total number of books read by the club for x months.

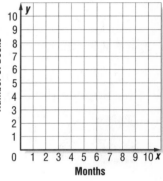

Write an equation. Let x represent the months and y represent the number of books.

_____

The slope is _____.

Vocabulary Check **Write the vocabulary word that completes each sentence.**

12 _____ is the ratio of the change in the y-value to the corresponding change in the x-value.

13 **Writing in Math** Explain how to find the slope for the equation $y = -2x + 7$.

_____

_____

_____

▶ Spiral Review

Graph another point on each line, given one point on the line and the slope. (Lesson 6-5, p. 248)

14

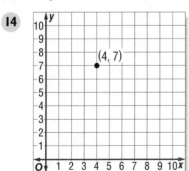

The slope is $-\dfrac{3}{4}$.

15

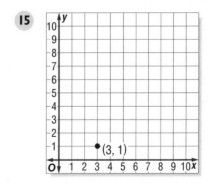

The slope is $\dfrac{3}{2}$.

STOP

Copyright © Glencoe/McGraw-Hill, a division of The McGraw-Hill Companies, Inc.

Find the slope of the line.

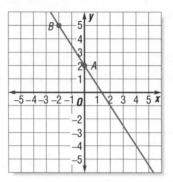

1 What is the "rise" of the line? _____

What is the "run" of the line? _____

What is the slope of the line? _____

Graph each equation.

2 Graph $y = 5x - 1$ and determine the slope of the line.

x	−1	0	1
y			

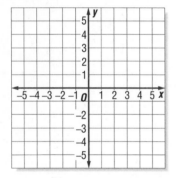

$$m = \frac{\boxed{} - \boxed{}}{\boxed{} - \boxed{}} = \frac{\boxed{}}{\boxed{}} = \frac{\boxed{}}{\boxed{}} = \underline{}$$

3 Graph $y = 3x$ and determine the slope of the line.

x	−2	−1	0	1	2
y					

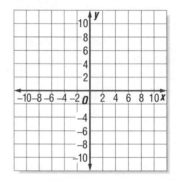

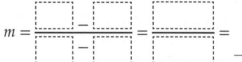

$$m = \frac{\boxed{} - \boxed{}}{\boxed{} - \boxed{}} = \frac{\boxed{}}{\boxed{}} = \underline{}$$

4 **MOVIES** The cost of a movie ticket is $8. Graph an equation that shows the cost of x tickets. Write an equation. Let x represent the number of tickets and y represent the cost.

_____

The slope is _____.

Ticket Prices

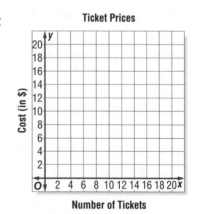

Copyright © Glencoe/McGraw-Hill, a division of The McGraw-Hill Companies, Inc.

Study Guide

Vocabulary and Concept Check

base, *p. 218*

coordinate grid, *p. 248*

estimate, *p. 232*

exponent, *p. 218*

factor, *p. 218*

hypotenuse, *p. 239*

inverse operations, *p. 225*

legs, *p. 239*

origin, *p. 248*

Pythagorean Theorem, *p. 239*

radical sign, *p. 225*

slope, *p. 248*

square of a number, *p. 218*

square root, *p. 225*

x-coordinate, *p. 248*

y-coordinate, *p. 248*

Write the vocabulary word that completes each sentence.

1. In a power, the number used as a factor is the _____.

2. _____ describes the rate of change between any two points on a line.

3. _____ are operations which undo each other, such as squaring a number and finding its square root.

4. The product of a number multiplied by itself is called the _____.

5. The _____ indicates a nonnegative square root.

6. The _____ can be used to find the length of the hypotenuse of a right triangle.

7. The _____ is the side of a right triangle that is opposite of the right angle.

8. A square number has a(n) _____ of 2.

Lesson Review

6-1 Squaring a Number (pp. 218–224)

Write an equation using exponents to represent the model.

9. _____ = _____

Evaluate each expression.

10. 2^2 _____

11. 7^2 _____

12. 4^2 _____

13. 9^2 _____

Example 1

Evaluate the expression 8^2.

1. What number is the base? **8**

2. What number is the exponent? **2**

3. Write the base the number of times given by the exponent. **8 · 8**

4. What is the value of the expression? **64**

6-2 Square Roots (pp. 225–230)

Find the positive square root using an area model.

14

$\sqrt{25} =$ _____

15

$\sqrt{16} =$ _____

16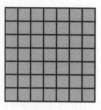

$\sqrt{49} =$ _____

Find the positive square root of each number.

17 169

$\sqrt{169} =$ _____

18 36

$\sqrt{36} =$ _____

19 81

$\sqrt{81} =$ _____

20 121

$\sqrt{121} =$ _____

21 196

$\sqrt{196} =$ _____

Example 2

Find the positive square root using an area model.

1. How many square units are shown? **4**

2. Are the number of rows and columns equal? **yes**

3. How many columns are shown? **2**

4. How many rows are shown? **2**

5. $\sqrt{4} = $ **2**

Example 3

Find the positive square root of 49.

1. Write the expression.

   $\sqrt{49}$

2. Name the factors of 49.

   1 · 49, 7 · 7

3. Replace 49 with the set of identical factors.

   $\sqrt{7 \cdot 7}$

4. $\sqrt{49} = $ **7**

Copyright © Glencoe/McGraw-Hill, a division of The McGraw-Hill Companies, Inc.

6-3 Approximate Square Roots (pp. 232–238)

Estimate each square root to the nearest whole number. Plot each value on a number line.

22 $\sqrt{10}$ is close to the whole number

_____.

23 $\sqrt{32}$ is close to the whole number

_____.

24 $\sqrt{119}$ is close to the whole number

_____.

Choose a reasonable estimate for each square root.

25 $\sqrt{102}$

    9.8   10.1   10.4

26 $\sqrt{167}$

    12.1   12.6   12.9

27 $\sqrt{198}$

    14.1   14.4   14.9

28 $\sqrt{213}$

    13.9   14.6   15.1

Example 4

Estimate $\sqrt{23}$ to the nearest whole number.

1. Write an inequality using common square roots.

$$\sqrt{16} < \sqrt{23} < \sqrt{25}$$

2. Find the values of the common square roots.

$$\sqrt{16} = \sqrt{4 \cdot 4} = 4 \qquad \sqrt{25} = \sqrt{5 \cdot 5} = 5$$

3. Plot the values of each square root on the number line.

4. So, $\sqrt{23}$ is between 4 and 5.
   Since $\sqrt{23}$ is closer to $\sqrt{25}$ than $\sqrt{16}$, $\sqrt{23}$ is closer to the whole number 5.

Example 5

Choose a reasonable estimate for $\sqrt{47}$.

1. Write an inequality using common square roots.

$$\sqrt{36} < \sqrt{47} < \sqrt{49}$$

2. So, $\sqrt{47}$ is between 6 and 7.

3. Circle the reasonable estimate.

   6.9   7.1   7.6

Copyright © Glencoe/McGraw-Hill, a division of The McGraw-Hill Companies, Inc.

6-4 Pythagorean Theorem (pp. 239–246)

Find the length of the leg of each right triangle.

29 $a = $ _____ units

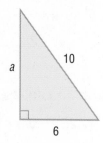

30 $b = $ _____ units

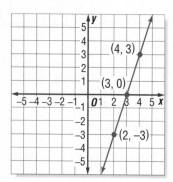

Example 6

Find the length of the leg of the right triangle.

1. From the figure, you know $a = 15$ and $c = 25$.

2. Substitute the values of a and c in the Pythagorean Theorem and solve for b.

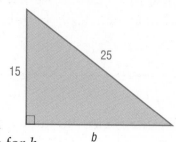

$$a^2 + b^2 = c^2$$
$$15^2 + b^2 = 25^2$$
$$225 + b^2 = 625$$
$$b^2 = 400$$
$$b = 20$$

3. The length of the leg is 20 units.

6-5 Introduction to Slope (pp. 248–254)

Find the slope of the line.

31

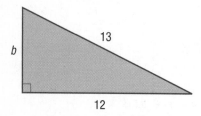

The rise is _____ units.

The run is _____ units.

The slope is _____.

$$\frac{\text{rise}}{\text{run}} = \frac{\boxed{}}{\boxed{}} = \underline{\quad\quad}$$

Example 7

Find the slope of the line.

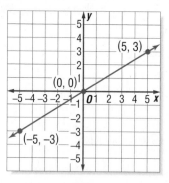

1. The rise is $+3$ units. The run is $+5$ units.

2. The slope is positive.

3. $\dfrac{\text{rise}}{\text{run}} = \dfrac{+3}{+5} = \dfrac{3}{5}$

Copyright © Glencoe/McGraw-Hill, a division of The McGraw-Hill Companies, Inc.

6-6 Slope (pp. 255–261)

32 Graph $y = 3x - 1$ and determine the slope of the line.

1. Complete a function table for the equation.

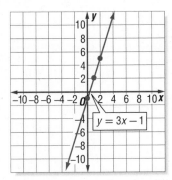

x	−2	−1	0	1	2
y					

2. Plot the points and draw the line on the graph.

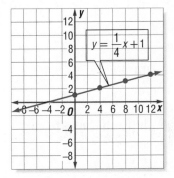

$y = 3x - 1$

3. Using two points on the graph, determine the slope.

$$m = \frac{\boxed{} - \boxed{}}{\boxed{} - \boxed{}}$$

$$= \frac{\boxed{}}{\boxed{}}$$

$$= \frac{\boxed{}}{\boxed{}} = \underline{}$$

Example 8

Graph $y = \frac{1}{4}x + 1$ and determine the slope of the line.

1. Complete a function table for the equation.

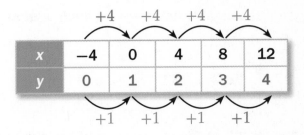

	+4	+4	+4	+4
x	−4	0	4	8	12
y	0	1	2	3	4

+1 +1 +1 +1

2. Plot the points and draw the line on the graph.

$y = \frac{1}{4}x + 1$

3. Using two points on the graph, determine the slope. **4**

$$m = \frac{1 - 0}{0 - (-4)} = \frac{1}{0 + 4} = \frac{1}{4}$$

Copyright © Glencoe/McGraw-Hill, a division of The McGraw-Hill Companies, Inc.

Evaluate each expression.

1. 4^2 _____

2. 7^2 _____

3. 11^2 _____

4. $\sqrt{64}$ _____

Choose a reasonable estimate for each square root.

5. $\sqrt{161}$

    12.2   12.7   13.1

6. $\sqrt{227}$

    14.8   15.1   15.6

Find the length of the hypotenuse of the right triangle.

7.

$c = $ _____ units

8.

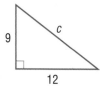

$c = $ _____ units

Determine the slope of the graph.

9. _____

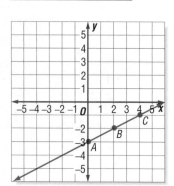

10. _____

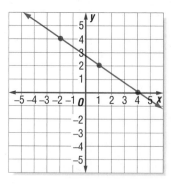

Estimate $\sqrt{96}$ to the nearest whole number. Plot each value on a number line.

11. $\sqrt{96}$ is close to the whole number _____.

Copyright © Glencoe/McGraw-Hill, a division of The McGraw-Hill Companies, Inc.

Graph each equation and determine the slope of the line.

12 $y = x + 2$

slope = _____

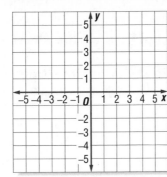

13 $y = -x$

slope = _____

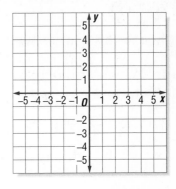

Solve.

14 **ART**   Mariah is mixing paint for an art project. She will make 4 gallons of green paint. Each gallon of green paint requires 3 quarts of yellow paint. How many quarts of yellow paint does she need to make x gallons of green paint?

$y =$ _____

Gallons of Green Paint, x	1	2	3	4
Quarts of Yellow Paint, y				

How many quarts of yellow paint would Mariah use to make all 4 gallons of green paint? _____

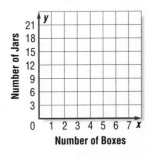

15 **PACKAGING**   Each box at The Cardboard Warehouse can hold 6 glass jars. Graph an equation to represent the total number of jars contained in x boxes.

The equation is _____.

The slope is _____.

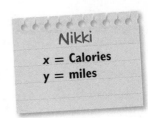

Correct the mistakes.

16 On Nikki's math quiz, the problem stated: "For every mile you bicycle, you burn 35 Calories. Fill in the blanks."

Let $x =$ _____; let $y =$ _____.

Nikki
$x =$ Calories
$y =$ miles

What is the mistake Nikki made?

_____

STOP

Copyright © Glencoe/McGraw-Hill, a division of The McGraw-Hill Companies, Inc.

Choose the best answer and fill in the corresponding circle on the sheet at right.

1 Which ordered pair falls on the line of the equation $y = x - 2$?

    A $(-2, -4)$ **C** $(-3, -2)$

    B $(0, -4)$ **D** $(3, 2)$

2 Maribel is tiling her kitchen floor. She sets the square tiles in 7 rows and 7 columns. How many tiles will Maribel use?

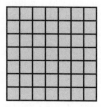

    A 36   **C** 52

    B 49   **D** 64

3 Which ordered pair falls on the line of the equation $y = 3x + 4$?

    A $(-1, -1)$ **C** $(-3, -5)$

    B $(3, 4)$ **D** $(5, 9)$

4 Use the Pythagorean Theorem to find the triangle's missing side length.

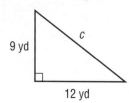

    A 11 yd   **C** 17 yd

    B 15 yd   **D** 18 yd

5 What is the slope of the line graphed below?

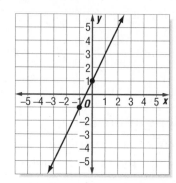

    A $-\dfrac{1}{2}$ **C** 1

    B $\dfrac{1}{2}$ **D** 2

6 What is the approximate positive square root of 7?

    A 2.6   **C** 4.9

    B 3.1   **D** 49

7 Jake is designing a square playground for his dog. It will cover an area of 64 square feet. What is the length and width of the playground?

    A 6 feet   **C** 8 feet

    B 7 feet   **D** 9 feet

Copyright © Glencoe/McGraw-Hill, a division of The McGraw-Hill Companies, Inc.

8 Find another point on the line that has a slope of $\frac{3}{4}$ and includes the point (0, 1).

A (−4, −3)   C (4, 3)

B (3, 4)   D (4, 4)

---

9 What is the slope of the line graphed below?

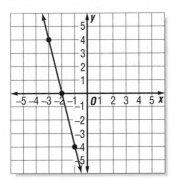

A $-\dfrac{1}{4}$ C 1

B $\dfrac{1}{4}$ D −4

---

10 Which point on the number line below represents $\sqrt{47}$?

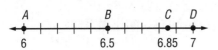

A Point *A*   C Point *C*

B Point *B*   D Point *D*

11 The length of the shadow of a light pole is 20 feet. The distance from the end of the shadow to the top of the light pole is 16 feet. How tall is the light pole?

A 12 feet   C 14 feet

B 13 feet   D 15 feet

ANSWER SHEET

Directions: Fill in the circle of each correct answer.

1 Ⓐ Ⓑ Ⓒ Ⓓ

2 Ⓐ Ⓑ Ⓒ Ⓓ

3 Ⓐ Ⓑ Ⓒ Ⓓ

4 Ⓐ Ⓑ Ⓒ Ⓓ

5 Ⓐ Ⓑ Ⓒ Ⓓ

6 Ⓐ Ⓑ Ⓒ Ⓓ

7 Ⓐ Ⓑ Ⓒ Ⓓ

8 Ⓐ Ⓑ Ⓒ Ⓓ

9 Ⓐ Ⓑ Ⓒ Ⓓ

10 Ⓐ Ⓑ Ⓒ Ⓓ

11 Ⓐ Ⓑ Ⓒ Ⓓ

Success Strategy

If you do not know the answer to a question, go on to the next question. Come back to the problem, if you have time. You might find another question later in the test that will help you figure out the skipped problem.

Copyright © Glencoe/McGraw-Hill, a division of The McGraw-Hill Companies, Inc.

Copyright © Glencoe/McGraw-Hill, a division of The McGraw-Hill Companies, Inc.

number, squaring a, 218–224

obtuse angles, 138, 145

origin, 248

parallel lines, 161

pattern, look for a, 165

Practice. See Step-by-Step Practice

Problem-Solving. See Step-by-Step Problem Solving

Problem-Solving Strategies, 142, 148, 157, 165, 184, 190, 198, 205, 221, 228, 236, 252, 259

Progress Check, 152, 168, 193, 208, 231, 247, 262

proportion, 194–200
   defined, 194, 201
   solve problems using, 201–207

protractor, 138

Pythagorean Theorem, 239–246

radical sign, 225

rate
   defined, 187
   and unit costs, 187–192

ratio, 180–186

ray, 138
   common, 153

Real-World Applications
   advertising, 245
   agriculture, 158
   animals, 183
   architecture, 151
   area, 238
   art, 150, 157, 213, 269
   baskets, 184
   biking, 198
   book club, 261

bridges, 167
business, 191, 199
canoeing, 261
chairs, 231
chess, 185, 230
class supplies, 230
clocks, 144
computers, 184
construction, 148, 175, 245
cooking, 175
cross-country, 159
distance, 246
doors, 175
earth science, 208
entertainment, 207
events, 206
farming, 224
fitness, 207
flags, 149, 247
flowers, 183
food, 199
football, 185
games, 186, 230, 238
gardening, 159
gardens, 228, 237
gas mileage, 200
geometry, 205
grades, 186
health, 221
home improvement, 244
interior design, 144, 221
kites, 142
labels, 175
ladders, 254
logos, 151
maps, 166
money, 260
movies, 262
music, 184
nature, 190
office space, 224
packaging, 269
patios, 236
pets, 159, 198
population, 191, 192, 200
prices, 259
puzzles, 223, 231
quilts, 165
railroads, 165
ramps, 143, 252
reading, 199
rectangles, 247
repairs, 168
reunion, 224
road signs, 142
safety, 207

sails, 254
sales, 192
school, 198
science, 222
sculpture, 159
shadows, 205, 254
skiing, 252
spelling, 193
sports, 158, 184, 193
stairs, 165
tents, 167
tiles, 149, 228
time, 152
travel, 208, 213, 246
video games, 238
work, 204
writing, 213

rectangles, real-world applications, 247

Reflect, 143, 149, 158, 165, 185, 191, 199, 206, 222, 229, 237, 245, 252, 260

right angles, 138, 145, 153

roots, square, 225–230

side, 145

similar figure, 201

simplest form, 180

Skills, Concepts, and Problem Solving, 143, 150, 158, 166, 185, 191, 199, 206, 222, 229, 237, 245, 253, 260

slope
   formula for, 255–261
   introduction to, 248–254

Spiral Review, 151, 160, 167, 192, 200, 207, 230, 238, 246, 254, 261

square of a number, 218

square roots, 225–230
   approximate, 232–238

squaring a number, 218–224

Step-by-Step Practice, 141, 147, 156, 164, 183, 189, 197, 204, 220, 227, 235, 243, 251, 258

Step-by-Step Problem-Solving Practice
   draw a diagram, 221, 244

Copyright © Glencoe/McGraw-Hill, a division of The McGraw-Hill Companies, Inc.

make a list, 184
make a table, 198
solve a similar problem, 190
use a diagram, 142, 148
use a formula, 252
use a model, 228
use a table, 205
use logical reasoning, 236
write an equation, 259

straight angles, 138, 153

Study Guide, 169–173, 209–211, 263–267

Success Strategy, 177, 215, 271

supplementary angles, 153

table
make a, 198
use a, 205

Test Practice, 176–177, 214–215, 270–271

transversals, 161–167
triangles, 145–151

unit costs
defined, 187
and rates, 187–192

unit rate, 187

vertex, 138

Vocabulary, 138, 145, 153, 161, 180, 187, 194, 201, 218, 225, 232, 239, 248, 255

Vocabulary Check, 144, 151, 160, 167, 186, 192, 200, 207, 224, 230, 246, 254, 261

Vocabulary and Concept Check, 169, 209, 263

Who is Correct?, 140, 147, 155, 163, 182, 188, 196, 203, 219, 226, 234, 242, 250, 257

Writing in Math, 144, 151, 160, 167, 186, 192, 200, 207, 224, 230, 246, 254, 261

x-coordinate, 248

y-coordinate, 248

Your Turn, 139–140, 146, 154–155, 162–163, 181–182, 187–188, 195–196, 202–203, 219, 226, 233–234, 240–242, 249, 256–257

Copyright © Glencoe/McGraw-Hill, a division of The McGraw-Hill Companies, Inc.